How to be a Better Scientist

T0139218

Understanding the fundamentals of conducting good science, that will have an impact, is the goal of every aspiring scientist. Providing a wealth of tips, *How to be a Better Scientist* is the book to read if you want to succeed in this competitive field.

Helping readers gain an insight into what good science means and how to conduct it, this book is ideal to read cover-to-cover or dip into. It includes easily accessible guidance on topics such as:

- What characteristics should a scientist have?
- Understanding the hypothesis
- Integrity in science
- Lack of confidence and the embarrassment factor
- Time management
- Coping with rejection
- Interacting with the science community.

With its broad focus, this friendly guide will enthuse, inspire and challenge, and is an essential companion for all aspiring scientists.

Andrew C. Johnson is a Principal Scientific Officer at the Centre for Ecology & Hydrology, UK, and a Visiting Professor at Brunel University London, UK.

John P. Sumpter, OBE, is a Professor of Ecotoxicology at Brunel University London, UK.

How to be a Better Scientist

**Andrew C. Johnson and
John P. Sumpter**

Routledge
Taylor & Francis Group

LONDON AND NEW YORK

First published 2019
by Routledge
2 Park Square, Milton Park, Abingdon, Oxon OX14 4RN

and by Routledge
52 Vanderbilt Avenue, New York, NY 10017

Routledge is an imprint of the Taylor & Francis Group, an informa business

British Library Cataloguing-in-Publication Data
A catalogue record for this book is available from the British Library

Library of Congress Cataloging-in-Publication Data
Names: Johnson, Andrew (Andrew C.), author. | Sumpter, John (John P.), author.
Title: How to be a better scientist : researching with impact / Andrew Johnson
    and John Sumpter.
Description: Abingdon, Oxon ; New York, NY : Routledge, 2019.
Identifiers: LCCN 2018032468| ISBN 9781138731219 (hardback) |
    ISBN 9781138731295 (pbk.) | ISBN 9781315189079 (ebk.)
Subjects: LCSH: Research—Methodology—Study and teaching (Graduate) |
    Science—Methodology—Study and teaching (Graduate)
Classification: LCC Q180.55.M4 J64 2019 | DDC 507.2/1—dc23
LC record available at https://lccn.loc.gov/2018032468

ISBN: 978-1-138-73121-9 (hbk)
ISBN: 978-1-138-73129-5 (pbk)
ISBN: 978-1-315-18907-9 (ebk)

Typeset in Celeste and Optima
by Swales &Willis Ltd, Exeter, Devon, UK

Contents

Contents

Acknowledgements

The authors are very grateful to Sophie Bishop whose artwork has done a lot to give our book the friendly feel we wanted. Johnson would like to thank his CEH colleagues, Mike Hutchins, Monika Juergens and Andrew Singer, with whom he's had many discussions over the years on what exactly constitutes good science, with the aid of tea and biscuits. Thanks are also due to Wolfgang Hofgartner for his expertise and to the many young researchers including Han Zhang, Anna Freeman, Holly Tipper, Qiong Liu, Steph Chaousis and Charlie Outhwaite, whose advice and encouragement greatly helped the authors. Johnson would like to acknowledge the support provided by his host institution of CEH. Sumpter would like to thank a number of his own students, especially Rumi Tanoue, who not only provided her own thoughts but also sought those of her Japanese colleagues and also other students based at Brunel University London. He also thanks his university for giving him the freedom to write this book.

Author biographies

Andrew Johnson started his PhD back in 1985 in the Soil Science Department of Reading University, UK. He had the good fortune to be supervised by someone who was a model scientist – Dr Martin Wood. The experience gave him the confidence to pursue a research career, but also stimulated his thinking on what distinguishes good science and scientists. After Reading University he enjoyed further good fortune in coming under the wing of the then Institute of Hydrology, now Centre for Ecology and Hydrology in Wallingford. He believes he caught them at a weak moment!

John Sumpter began his PhD in 1973 in the Department of Zoology of Bangor University, North Wales. He had two superb supervisors, Professor Jimmie Dodd and Professor Brian Follett, although it was not until he became a supervisor himself that he realised how good they were, and how much he had learnt from them. He has subsequently supervised over 50 of his own students while they did their PhDs and post-doctoral research. While doing so he learnt much more from them than they realised. He has tried to incorporate into this book all that he has learnt about what it is like to be a young scientist learning his or her trade.

Foreword

For those wishing to improve their knowledge about the achievements of science there are thousands of books on every discipline under the sun. There are also less specialised, populist science books which bring their subject to life for a wide audience. The scientists wishing to improve their technical skills, be it in practical matters such as analytical chemistry, or general skills such as writing scientific publications, will find their needs catered for. We are able to benefit from memoirs and advice from past and present eminent scientists. There are an increasing number of more technical books which prepare, inform and guide PhD students through the typical experience in a Western country.

But this book has another focus. It is to make you a better scientist. That is, to give you the best chance of having a successful and rewarding career in science. Thus, we primarily want you to understand the scientific method, so that you do better experiments, and to make you a better communicator. This understanding should be relevant to you in whatever branch of science you choose, or the environment you work in, be it academic, government, industrial or otherwise. In due course it is likely that you too will become a supervisor and then science manager, so we also offer advice to prepare you for these roles. Along the way we will help you with the social and human aspects of the scientific career too. So we hope this book will be part of your training as a scientist. It tackles many of the issues and problems you will encounter, and suggest ways to address them. It is, in essence, a practical guide, aimed at helping you develop into the very best scientist you are capable of becoming.

What was our motivation in writing this book? The authors have supervised many PhD candidates, guided many post-docs and worked with countless scientists. But we have witnessed poor supervision and guidance of scientists, which have sometimes led to them seeking alternative careers, and there was nothing we could do about it. We have also observed older, and in theory more experienced, scientists doing poor science. We have enjoyed working with young scientists from many countries who are desperate to learn, yet whose resources and guidance can be painfully lacking. Many of these young scientists have a good grounding in their subject and are technically competent. But such knowledge does not guarantee that they will go on to become good scientists. We witness poor or indifferent science in the scientific publications and grant proposals that come across our desk every day. Yet from time to time our day has been brightened by clear, crisp papers describing excellent science. We want you to be one of these latter authors and not the former! Whilst scientists doing their PhDs and post-docs are at the stage of their career when they are most eager to learn and might benefit most from this book, we do not wish to focus exclusively on them. We hope the book might interest and intrigue those at an earlier stage who are considering a career in science, either in school, or doing a first degree or MSc. But we believe the guidance could help scientists at any stage of their career. Scientists should never consider themselves 'the finished article', more 'a work in progress'. We entertain some hope that those outside science, who nevertheless use scientific evidence, such as those in in business, politics, journalism or the law, will gain insights they might find helpful. Nor do we wish to put off the merely curious, the man or woman in the street who asks what is this thing called science, how is it different from other human activities?

We will take you from high concepts of the philosophy of science to more prosaic advice on tweaking your presentation and the management of your emotions. You may think the range of advice here has some odd combinations, but our aim is that the book will be the helpful friend in your pocket, a friend able to field the range of questions you might have as you move towards becoming a better scientist.

The world needs science and, to progress, must be informed by sound scientific evidence. This means all nations will benefit from a supply of qualified and thoughtful scientists. More people doing and understanding good science offers the greatest hope we have for humanity's future.

CHAPTER 1

What do we mean by science and 'being a better scientist'?

Scientists and the scientific method

If there is one chapter that could be judged as superfluous to this book, surely it should be this one! After all, we all know what science is and indeed what a good scientist looks like, don't we? But when we asked a range of colleagues 'what is science', we realised how vague the essentials are to both new and experienced scientists. We invite you to ask those around you how science works? Or to put it another way – what distinguishes the scientific approach from other human activities? Answers we have heard include; 'being logical, searching for truths, an orderly arrangement of facts and identifying nature's laws. But these are insufficient because they don't clearly distinguish what is special about the scientific method. So what might distinguish scientists and the scientific method from others?

Proposing a hypothesis

Most groups, clubs, institutions and individuals hold cherished beliefs through which they view the world. A scientist has a willingness to test a belief against the evidence. The scientist gets to the heart of an issue by converting those swirling ideas and heated discussions into a testable statement called 'the hypothesis'.

Gathering evidence

Everyone can have an opinion but scientists have a duty to evaluate issues through evidence. The wider and more comprehensive the information gathering (evidence collection) the stronger the foundations of any judgement that might ensue. To paraphrase one of the great scientists of the twentieth century, Lord Kelvin, 'if you can't measure it, you can't talk about it!' The evidence gathered from the environment or our experiments is then used to test (challenge) our hypothesis.

Dealing with uncertainty

Unfortunately evidence can be (and often is) contradictory. This is where science parts company from many other human responses on controversial issues. The scientific duty is to support the case with the largest body (weight) of evidence. This may be uncomfortable. The unswervable loyalty of the scientist must be to the data, the evidence.

Evidence and science are commonly misused in two ways. First, some choose to 'cherry pick' isolated evidence that supports their convictions whilst conveniently ignoring information which does not. Second, some may attempt to avoid taking action by pointing out the lack of complete scientific consensus and thus promote doubts. Yes, science involves argument and uncertainty but, where action is required, we must follow the weight of evidence.

Coming to a tentative conclusion

There is often a temptation when looking at the evidence to leap to the 'obvious conclusion' – that somehow the data support a long cherished belief of ours. But be careful! We must always review as many possible alternative explanations as we can. Then we support the case where the greatest evidence lies but always admitting to a varying degree of uncertainty. That is why we do not use the term 'proved'. We offer support but must acknowledge that further data may yet overturn our conclusion. With our own experiments, we design and describe them carefully, with the aim that someone else could repeat them and come to the same result. As more and more evidence supports a hypothesis, we must acknowledge that it becomes stronger. But this is still not truth or validation. Ironically, our only certainties lie in what can be disproved.

We would strongly recommend you avoid talking of things being proved, as being facts or as having been validated. These statements imply that no further challenges are necessary or indeed acceptable. That we are certain no new evidence will ever emerge to disprove our theory. But this would be a fundamental misunderstanding of science. We should accept the possibility, however remote, that in due course things may be overturned and replaced by a better theory. We will discuss the philosophy behind this in Chapter 3 on the hypothesis. Thus, although many of us leave school believing that scientific knowledge is irrefutable fact, it is better to view the world the other way round, that a small proportion is not going to be overturned (e.g. the heart is the organ which pumps blood round the body, the earth orbits the sun), whilst most will in due course be refined, altered or completely overturned.

Some have criticised science as being just another belief that is competing for an audience. In other words it should be given no greater credence than other strongly held opinions. This is wrong; science is not dependent on faith or conviction. It is really a philosophical approach to life's many questions that submits its (tentative) conclusions continually to test and scrutiny. It is therefore a method whose results and conclusions are continually open to correction. As we shall discuss later, for science to function it needs its practitioners to maintain objectivity. By maintaining objectivity a scientist has a vital function for society. This objective approach to a problem makes science unique.

Now you can see why scientists can be unpopular! First of all we have the temerity to ask questions about cherished beliefs. Then, if we are asked to respond to a question, we don't make a snap judgement, we take our time compiling and evaluating evidence, then when we finally produce an answer it is not unequivocal. Unfortunately, it is easy to be tempted by journalists, lawyers and politicians or perhaps by family and friends to give answers to scientific questions with resounding certainty. After all, we feel flattered to be asked our opinion, perhaps we want to please or even appear a superior being, dispensing the 'truth'. Good scientists will always admit that some uncertainty must exist in their answers. Despite getting a 'bad press', it is hard to see how modern society could function without us!

Science is more than an activity

When you first enter a scientific establishment you may simply have to follow the orders and instructions of a superior and perhaps adhere to a rigid protocol. Whilst the experience you gain and support you provide is helpful to any scientific endeavour, the budding scientist must think hard about what is going on and why? If you don't, science will be just another routine activity from which you won't get much pleasure. If you remain dulled by this experience and fail to rigorously question what you are doing and why, your work will be uninspired, and your research papers won't get much attention.

To many, putting on a white lab coat and carrying out a scientific activity which generates lots of 'interesting' data is sufficient in itself. In other words, science to them is merely a craft-like activity that helps to pass the time and pay the bills. If you follow more or less what everyone else seems to be doing, surely you can't go wrong? Perhaps not, but science won't move forward that much either!

What do we mean by 'a good scientist'?

Is it about financial success or esteem?

A good scientist, as judged by society and a parent institution, may well be someone who succeeds in continually getting funding and support. Someone who is on many committees and perhaps even on television. All of this may be valuable . . . But this is not necessarily the same as being a scientist who has driven science and human knowledge forward to the greater benefit of society. We can evaluate research scientists through their publications, which provide a window on their efforts. Their views are there to be challenged. Many scientists are employed in business, government and charitable organisations. The esteem they can enjoy in their organisations will be from the balance and objectivity consistently present in their advice. They will be acknowledged for their courage in sticking to the data, or evidence, however uncomfortable that may be.

Because we care about science and its lasting value to humanity, our recommendations in this book will focus on preparing you to be the scientist capable of genuinely moving science and society forward.

How would I spot a good scientist in my department?

There was an ironic statement made a while ago that you could judge the prestige of a scientist by how long they held up progress in their field! In other words this is someone who jealously defended their own view and ruthlessly crushed alternative ideas, perhaps someone who spends time only with other grand and superior scientists, someone who is too busy or high and mighty to take an interest in you? These qualities are the reverse of a good scientist. So how might you recognise a good scientist walking down your corridor?

Humility

Some scientists can become very grand. As a world expert they look down on lessor mortals and strike down those who contradict them. They are convinced of the power of their logic and they advertise their superior intelligence. However, the wise scientist is also humble. First, they know they depend on the support and collaboration of others. Second, they are aware of their own ignorance and the possibility that their own theories may be overturned at any time. They are ready and indeed keen to always listen and learn. A senior and respected scientist we both admire recently wrote a reflective paper in which he stated '90% of my ideas turn out to be wrong!'

Courage to challenge established thinking

They do not succumb to 'group think' and may be seen as something of an awkward colleague as they insist on evidence before giving support. It is a natural tendency to want to support and buttress the thinking of the community, perhaps always assuming that your professor must invariably be right simply because that person is the Professor and your boss. But this attitude actually does science and your professor no favours. By not being shackled to some dogma, the professor could put forward alternative visions that could drive science forward. He or she may know that there are bigger or more worthwhile challenges out there than those currently preoccupying their community or company.

Able to see the big picture

Whilst science is about getting the details right and squeezing out errors, there is a danger that by becoming immersed in detail, scientists can lose sight of the bigger picture. They can lose the ability to recognise what is most important and focus on, or at least acknowledge, the biggest and most rewarding issue in their field. Seeing the big picture could also be viewed as getting less worried about trivial issues and mistakes, rising above the minor mishaps and frictions and keeping sight of the purpose of the enterprise. This could be seen as being able to distinguish between the scientific process and the purpose of scientific endeavour.

Accepts with good grace the falsification of their own hypothesis

As we shall see, objectivity is a key characteristic in a scientist. One of the most formative experiences in the scientific career of one of this book's authors was the response of a PhD supervisor to work that disproved a hypothesis of his, one that he had been advocating in several papers over a number of years. To my surprise, after a short consideration, he readily agreed that the data disproved his hypothesis and heartily congratulated me. In that instant I realised what a great scientist really was!

Honest

It is not unusual for scientists to fall prey to exaggeration about the implications of their results. In the bid to attract attention, citations and, hopefully, funding, the desire to hype up results may be overwhelming. In addition, some scientists may deliberately underplay or hide parts of their own data which inconveniently contradict 'the story'. This is bad for science and a very poor example to give to colleagues, particularly to the students who are working for these scientists. We shall discuss this problem in more detail in Chapter 5. An interesting dilemma here can be in the reporting of negative results. Unfortunately, they are not attractive to editors of journals and do not garner news headlines. However, they are much more valuable to science than at first sight and they also signal to the community that the individual is honest and can be trusted.

Makes time to encourage and teach

With the pressure to carry out research and maintain all the appearances of being a leading scientist, there are many excuses for failing to spend time teaching or passing on experience to junior scientists. Yet one of the best legacies to leave is the new scientist you have trained to think and practise good science. They do you credit but, more importantly, carry the torch themselves and pass it on to others on how best to do science.

Ability to communicate in a way that is easy to understand

Scientific techniques and equipment are complicated. The issues are complex and multi-faceted, whilst the results may be ambiguous. This complexity can lead to different disciplines developing their own unique language. It may seem clever to then communicate in a complex way to apparently demonstrate your own higher intelligence. Yet you will find that really good scientists are able to convey complex information in an extraordinarily clear and simple way. This clarity draws other people to work with them but, more importantly, it means they can disseminate their knowledge to the widest possible audience.

Permits open debate and encourages alternative views

For science to evolve and weak theories to be discarded, they need to be open to challenge. No one has a monopoly on wisdom, so the good scientist will invite ideas and challenges from all colleagues, high and low. This to and fro of ideas can lead to new insights. It is better that ideas are rigorously challenged and revised before they and their author are tripped up at an embarrassingly late stage, such as at an international conference! Questions from the most junior scientists should be warmly encouraged for two reasons. First, a new outlook may spot a mistake you haven't considered and second, if they are to develop into good scientists, they must get into the questioning habit early.

Welcomes collaboration

Some scientists who have worked for many years in a particular field and perhaps built up a reputation can feel the need to jealously guard it from interlopers. New entrants to the field are seen as rivals and offers to collaborate may be rebuffed. This is not good for the individual or their field of science. Instead, the good scientist is happy to proclaim their ignorance of fields or skills outside their own and seek out those who have complementary strengths. Knowing that scientific problems are often multi-faceted, they draw in and welcome these other scientists with different perspectives. Those who have confidence issues, or are selfish or jealous, are less likely to take this approach. Seeking additional support to tackle a science problem is not a weakness.

Successfully converts funding into outputs that reveal all these characteristics!

Science only exists in a concrete sense through its outputs. Science is more than carrying out an experiment well and generating data – this still has to be interpreted and converted into a product. This then must be communicated outside the lab such as in a paper, report or patent! Although scientific papers are expected to be written in the unemotional third person style, we find that the human characteristics described above can often be divined within them

But what about scientific genius?

It must be admitted that occasionally the scientific community is blessed with a person of genius. This is not something that can be taught! Also we accept that we may have to make exceptions to allow such a person to flourish.

This book is not about scientific genius. We look on these people with wonder but should not despair about ourselves. The majority of scientific advances in every field are based on ordinary scientists applying the scientific method and that person can be you!

That the characteristics above constitute a 'good scientist' are our personal opinions. We do not say they are all vital attributes of a good scientist but we would expect to see many of them present.

When looking at these characteristics of a good scientist they could also be seen as those of a 'good person'. If the virtues we will discuss in this book are understood and followed, they could benefit your life in general. Good scientists have much to commend them!

Checklist for Chapter 1: What do we mean by science and 'being a better scientist'?

---

1 You should view science as a method or way of thinking to address problems.

2 You are someone who is willing to propose a hypothesis.

3 You diligently gather evidence.

4 You give your support where the greatest weight of evidence lies.

5 You accept that some uncertainty remains.

6 You acknowledge that a conclusion is tentative.

7 You do not cherry pick from the data to argue your cause.

CHAPTER 2

What characteristics should I have as a scientist and am I that person?

As you leave school and then complete your Bachelor's degree there are many different careers that the world has to offer. If you persist and continue to produce good outputs, science is likely to provide steady employment. In industry there are increasing possibilities to be very well rewarded, particularly where science innovation is a key part of the business. Where science differs from many other professions is in the satisfaction you will have of knowing that it is valued by society, that it asks you to use your creativity and innovate as an individual and that, thanks to your papers, you may soon become known and recognised worldwide by your peers. Science is truly international and researchers will read and follow the best science regardless of nationality.

Don't worry! No one person ever has the perfect set of all the right skills or characteristics desirable in a scientist and it would be foolish to think so. However, many people have the potential to become scientists. In this book we will try to guide you towards becoming a good scientist. Many people are capable of learning sufficient technical skills to operate within science. But with a few extra skills they can produce outputs in their own right and become better scientists. It is of course helpful to be of brilliant intelligence but this is not vital and indeed quite rare. Those who have the capacity to improve as scientists are distinguished by the way they assess and respond to data/ information/evidence. When looking at data, from your own or others' experiments, the first rule must always be to 'clear your mind'! The data should be examined without prejudice, pre-conceived ideas, hopes or fears. You should not try to twist the data and interpretation to suit your own or your superior's pet theories. If you can look at the data in an open and honest way then you are 'the right person', you have the right stuff!

Finally, find and do something you like! Where in science are you keen, curious, excited and passionate? Finding a position on a topic you really care about will make the hard work bearable and worthwhile. We will discuss later how to find good supervisors or collaborators to help you on your path.

The character question

There are many characteristics that are desirable. They include being naturally curious, persistent, able to work on your own and in teams, meticulous and imaginative. It is helpful to be patient, passionate, focused, and a careful planner. Also it is valuable to be a good and sympathetic communicator on both paper and in oral presentations. Let's have a look at these in a bit more detail and decide which one is the most important:

Curiosity

Of this list being naturally curious is perhaps the most precious. Good science starts with that spark of curiosity, the 'I wonder why' question. This curiosity is what clearly distinguishes the first great scientists in human history. But you might feel that you are simply a small cog in the machine of a big department, who responds to orders from above. Whilst you may work in a large team with several managers above you, please do not let this feeling extinguish your precious curiosity! Be bold; keep asking those questions. It may appear we know nearly everything and the world has little need of yet another young researcher. But the wiser the scientist, the greater their acknowledgement of how little we know and the more we need young, fresh curious minds like yours!

Imagination

Imagination is as valued in science as in art, where you have the confidence of thinking, dreaming and trying something different. Without imagination we will not make progress. You could equate this with freedom to think and question in your own right. The great thing about imagination is that no person or institution has a monopoly. Anyone can deploy their imagination. You do not have to rigidly follow any pre-existing doctrine or method. You might come up with a new idea in the bath or at home whilst making some pasta! It could come from nowhere or be linked to something you observed in another field of science. Ultimately you will have to formalise your new theory/idea into a hypothesis, which we will discuss in the next chapter.

Willingness to learn

There are two important aspects to learning and character. The first is about having confidence and the second about avoiding arrogance. In the first case you will need from time to time to move to different aspects of science or simply extend your knowledge to help your existing project. This means taking your first steps in a new discipline. Surprisingly, many scientists never have the confidence to do this and remain steadfastly researching the same narrow topic throughout their lives. This can have unfortunate consequences when the funding runs out, or when the scientist needs to apply for a new position. Probably, what holds people back is that they dislike the vulnerability of going back from being the expert to the learner and potentially revealing their ignorance to others. You will need to overcome this and enjoy expanding your knowledge. The more widely you develop your knowledge, the better your science will be and the more employable you will become. Scientists benefit from humility, acknowledging their weaknesses and always desiring to expand their knowledge. It is best to say scientists will always be trying to find ways to improve throughout their careers.

So the second aspect is avoiding arrogance. In some scientists there can be a tendency to believe they 'know it all'. This can naturally develop with age or even be the case with precocious young scientists. The older scientist may feel they have 'served their apprenticeship', written several good papers and now it is time to take their place amongst the elite. This is a very dangerous tendency in the scientist, as it can lead to making sweeping pronouncements which turn out to be false. Such scientists do not develop: they are standing still and they are usually annoying their colleagues with their arrogance.

Able to plan

Good planning is central to successfully completing a project. In many ways, preparing for experiments is like preparing for a military operation! You attempt to leave nothing to chance. You may have to purchase the right equipment way ahead of time, think of the number of replicates and controls, get advice from statisticians, find the right window to use the facilities when they are free and find out the right time for someone who is already experienced to advise you. Quite possibly you will be working with more than one colleague on your project, so you will need good team work to succeed. This might include delegating different tasks according to your colleagues' skills and availability. You have to think more widely to consider how long the experiment will take and how many you can do in the lifetime of the project? Then, as is all too often the case, you may have to adapt your planning when an experiment fails. Writing up your work or pulling it together into a report or paper takes much more time than you think. But you must always remember that time is finite. Do not strive for an unreachable perfection when what you have is perfectly acceptable and will do the job and do it well. You must concentrate and think clearly to allow yourself to deliver on time. There is nothing that makes a supervisor or collaborator happier than receiving that piece of work from you on time!

Meticulous in their work

You have to care! The experiment, report, paper or patent is your finished product. It has your name stamped on it. This is where it pays to really think carefully about the design of an experiment and also about the execution. It always helps to have a clean, tidy and organised work space. Everything you need in your experiment should be clearly labelled. Look really carefully at the data. Does it all make sense? Have there been some errors in the calculation step? Learn from your mistakes to ensure things go better the next time!

Persistent

It is very rare in science for things to come easily! If you are the sort of person who gives up easily then this career may not suit you. My first supervisor told me that even with the best scientists, one in three experiments end in complete failure. But sometimes even a success of two out of three is on the optimistic side. It may take months, or even years, to get a key experiment to work following repeated failures. This is a price that has to be paid when you work at the cutting edge of science. You will have to be methodical as, step by step, you learn where things are going wrong and then adapt your approach. Your reward will be all the sweeter when you finally get things to work. This hard road of experience is something you never forget. Later on in your career that knowledge will help you to advise others in the same predicament.

Not afraid of hard work

Persistence and meticulousness are typical components of working hard. The ability to work to a high standard to meet a tight deadline can mean having to work long and sometimes unsocial hours. Producing something to world class standards, such as a good science paper, at a time when the information is most needed by the wider community, will entail sacrifices. There will be times when a significant effort will be needed, and you will have to forgo meeting your friends in the bar for a period of time. But what makes this palatable is that, hopefully, your name will be on that paper and your effort recognised in perpetuity. However, we are not at all suggesting that continual very long working hours will be good for you or for the quality of your work. We will discuss what supervisors can or should not expect from you in a later chapter.

Able to work alone

In our early days at school and then University we are always part of a community following a set of instructions. We don't really need to think, just act. But as a scientist you, will be given a problem and told to go away and solve it. To some this is wonderful moment where they are at last free to think and come up with approaches on their own. To others these first moments might provoke feelings of panic and fear. But don't despair! There is a wealth of knowledge you can access at your fingertips with search engines and publications online. There will be others in your department who may have worked in this area before and can advise you. You can then weigh up the different options and approaches that might lead to success. But the essential point is that there will be moments when you will be on your own and will have to use your own judgement. There is something beautiful in the tranquillity of deciding your own fate and direction. It can be comforting to be working away in the lab on your own, perhaps listening to music and doing your own thing. No one is bothering you with instructions and requests every five minutes; you leave your email behind and just do your own thing at your own pace.

Happy to work in a team

Although this may appear at first contradictory to having the confidence to work on your own, as you go on in science you will find that most scientific endeavours involve some level of team work. Scientific problems are usually very complex and require people with different expertise to come together. A number of skills will be required from you, but these will be largely in the social and emotional intelligence field. Make the product you deliver to the team of the highest quality possible (despite your feeling that there are other projects more important to you). Be as explicit as possible in explaining the data or report on your piece of the work: do not take any knowledge for granted. As you may be a project or task leader, everyone should be viewed as equals with important views to impart. Diplomacy will be essential, so do not show feelings of irritation or anger with the progress of your colleagues. Seek to find areas of agreement and common ground. Thus, the approach taken may not be the ideal one in your eyes but nevertheless it is workable. Listen respectfully to your colleagues and find ways of supporting them as well as thanking them for their contributions. You must yourself meet agreed deadlines, whilst using tact and diplomacy to encourage others to meet theirs. Respond as quickly as you can to emails; there is nothing that more strongly suggests you don't care about your team than being slow to get back to your colleagues' requests.

Focused

It is all very well to come up with a great plan but there will be many distractions that can blow you off course. These could be personal, to do with your home life or health, at work with issues with your colleagues, new administrative tasks to fulfil, requests to review papers or go to conferences, or new demands coming down from your supervisor. Smartphones are now a particular problem for many; their bleeps and vibrations can be irresistible. Your personal and home life must be attended to, of course, but hopefully without losing focus on your work goal. All those extra work tasks claiming your attention will each have some merit but, crucially, not equal merit! The essential point here is that if you don't complete your experiment, report or paper, then it won't get done. Both your career and your collaborators/supervisors are relying on you to complete that piece of work. They will not be impressed if you are behind schedule due to being distracted by relatively trivial requests. This may seem cold hearted, but there are literally 101 reasons for you not to complete your work. Draw up a plan but make sure that your key work is Number 1, or as close to Number 1 as possible. If you are really conflicted, then share your problem with your supervisor; be honest and let your supervisor guide you. Often, he or she can deflect these distractions away from you. You will find more advice on time management in Chapter 8.

A desire to communicate

It may be that you and a few others in your lab know you have just completed some excellent science. But until you communicate what you have done and why, your work could be overlooked. So it is highly desirable that you convert your enthusiasm for your topic into an explanation to others. However, communicating or explaining science to others is not easy. The rationale for the study may not be immediately apparent; the science may be complex and the results not easy to interpret. Whilst you have the advantage of having eaten, slept and breathed the project for months, if not years, this is not true for your potential audience. So you must have the desire to explain what you are doing and why it is important. The skill is in doing this using as clear and simple a language as possible, something we will examine later. But remember you must advertise and explain what you have done, or are doing, at every opportunity. Best of all, you will get feedback to improve your work and maybe even some praise, if you are lucky, that will help keep you going in those difficult days when your studies are not going so well! The various ways of communicating your science are discussed in Chapters 9 and 10.

However, there is only one quality that is **absolutely essential** in a scientist:

OBJECTIVITY!

The ability to be objective sets a scientist aside from most of the rest of society. The ability to be objective is why a scientist is vital to society; without objectivity, the scientist is just another lobbyist peddling a special interest. Unfortunately, it is not at all natural (human) to be objective. Our ego makes us passionate believers in our theory (I know it is right), sceptical of our opponents and believers in our friends. For example, we might readily believe information from an independent not-for-profit charity whilst remain disbelieving of information from industry or Government. We might cheerfully accept the data generated by expensive and state of the art analytical equipment and disdain contrary evidence from back of the envelope calculations. We might give credence to information that confirms the status quo, whilst dismissing those contradictory pieces of evidence. We might happily accept measurements reported from a respected lab in a developed nation but treat with suspicion those from a newcomer based in a developing nation. This stew of prejudice may be hidden, but is often close to the surface. Scientists are often quick to spot subjectivity in others but slow to see it in themselves. Social scientists have studied this phenomenon for a long time: it is called confirmation bias.

In a particularly bad case I knew of, a PhD student (we'll call him Gary and meet him again later) dismissed the results of any experiment of his own which disagreed with his theory. It was the fault of the experiment in some way. It was the experiment that had failed and not his theory. He had failed the objectivity test and so without further tutoring he could not hope to improve as a scientist.

Checklist for Chapter 2: What characteristics should I have as a scientist and am I that person?

---

1   Science is valued by society and essential for human progress. The training and mind-set you gain would be suitable for many types of employment.

2   Objectivity is the most vital of all characteristics of a scientist!

3   Be curious.

4   Imagination is helpful.

5   Keep learning throughout your career and try new things.

6   A desire to communicate.

7   Other valuable characteristics include being good at planning, meticulous, persistent, able to work hard, work alone or in teams, and stay focused.

CHAPTER 3

Understanding the hypothesis

Many years ago, a highly knowledgeable student I was supervising, yes it's Gary again, would charge off with great confidence and carry out a series of costly experiments. In the evening he would come to me with a sense of satisfaction and produce, with a flourish, his graphs. He was delighted to reveal to me that he had produced some 'very interesting results'. He would then head off home to an untroubled sleep, knowing he had done his job. After all, only very clever scientists at the top of their game were capable of producing 'very interesting results'. To him, science was an activity that piled up one set of interesting results upon another. And indeed it is often said that scientific research throws out more questions than answers. However, we in our project were not making tangible progress and certainly did not know any more than we did the previous month. The only certainty was that we were spending our way through the research money at an impressive rate. Each new interesting result was a siren voice inviting the student to spin off down ever more 'interesting' avenues. Soon, the lack of both money and time would bring the party to an end. Science had not moved forward, nor, to be frank, had it become more interesting! Many efforts were required to get Gary back on the right path.

You probably will have sensed yourself that there are many different avenues for science to explore, and many different experimental techniques to apply. You would love to play around with all of them, and indeed many different combinations of new ideas and novel techniques will yield 'interesting data'. But there are two problems here: 1) you won't have enough time and 2) interesting data are plentiful but they won't in themselves advance science.

Hypothesis to the rescue!

But what is a hypothesis, exactly?

Many confuse the hypothesis with an objective, an aim, an idea, or a theory but none of these accurately describes it. A hypothesis is a concise statement that is capable of falsification (or to put it another way, capable of being tested and disproved). It was the philosopher Karl Popper who insisted that the hypothesis was central to science. A topic that was not science, a pseudo-science, of which he gave astrology and Freudian psycho-analysis as examples, was one that could not offer a falsifiable hypothesis.

All students studying or doing science should have a (preferably single) hypothesis behind their research. They should be able to readily articulate this hypothesis. Their research should be aimed at testing this hypothesis. More precisely, what we mean by testing it is being able to falsify it. This applies whatever their topic; for example, it applies equally whether they are trying to determine if the earth is flat or round, or to design a more efficient car engine!

Let us try proposing some hypotheses that look as if we could falsify them:

1. Atoms are the smallest particles on earth.
2. All diseases are caused by viruses.
3. Steel is the strongest of all materials.
4. People over 60 years old cannot learn anything new.

5. Global warming is a natural phenomenon.
6. Health and safety in the workplace has reduced the number of accidents and injuries to people.

These are not all perfect hypotheses, but they will serve our purpose here. Thus, when we look at these hypotheses, our duty is to find at least one contradictory piece of evidence to falsify each statement. Only one contradictory example is needed to falsify a hypothesis. Without needing to do further experiments ourselves, we could find contradictory evidence that already exists in the literature for hypotheses 1–4. Unfortunately, hypothesis 5 does not lend itself to a falsifiable test, but the weight of evidence would strongly indicate an unnatural origin. With hypothesis 6, it seems feasible that we could find a weight of evidence to support this, although one could imagine the existence of some contradictory examples.

Here are some poor or inadequate hypotheses:

1. There is a relationship between feeling stressed and doing poor research.
2. Antibiotic resistance in rivers is linked to wastewater effluent.
3. Most flightless birds may be in danger of extinction.
4. Exposure to chemicals could be important to breast cancer.
5. It is likely water exists on some other planets.

Can you see how these hypotheses look rather difficult to falsify? How do we falsify 'a relationship' which could mean different things to different people? How do we address something that 'may play a role'? In these examples you are given an impression of what interests the proposer, yet the science which is needed is still unclear. Unfortunately, when we talk about relationships and linkages, this is all rather vague; it is not clear how to test the hypothesis and how much benefit we would get from doing so. So let us try to improve them:

1. Where a subject's cortisol (stress hormone) level and heart rate are 90% above the normal resting level they will perform poorly in a time-limited standardised test.

2. The range and quantity of antibiotic resistance genes found in river bed-sediments are directly proportional to the % wastewater in the river.

3. Numbers of Kiwi and Emu birds are now 10% of those in 1990.

4. All forms of breast cancer are linked to having a blood level of 50 ng/mL or greater of flame retardant chemicals.

5. All planets will have water present within 500 m of the surface.

You can see that the hypothesis is an absolute statement capable of falsification. It is a target we have put on the wall to aim at. To be useful it starts by being as broad in its scope as possible. It is also inviting us to take aim with our scientific rifle. Can we disprove it? If supported, such a hypothesis would be genuinely valuable to a wide range of people. Actually, it is providing us with several benefits. It clarifies what it is you want and need to do. You have a real, fixed target to aim at. It is triggering the type of experiment you need to design. As these hypotheses are falsified you will, from an early stage, be considering more specific hypotheses that might be more plausible.

But you will say that the world is very complex – surely a combination of several factors influences an outcome? Yes, but usually there will only be a couple of strong drivers and several weaker or irrelevant ones. With the hypothesis, we can eliminate the weaker ones. Another advantage of a clear hypothesis is that it will substantially reduce the chance of your experiments producing results that cannot be repeated by other scientists (see Chapter 7: The reproducibility crisis).

Fun examples of forming a hypothesis and designing suitable tests

You and your friends are arguing as to whether some chocolate in the afternoon actually helps your work performance. As the argument goes on, you propose a hypothesis:

The performance of a student who eats a 200 Calorie chocolate snack containing 60% cacao will exceed that of someone who doesn't eat this snack.

So now you have to design the test to falsify this hypothesis. The more subjects you enrol the more powerful the test. The wider the range of subjects, their age, sex and ethnicity, the better. Then you will consider the controls, one clearly is no snack, a second type of control might be a

200 Calorie snack that contains no chocolate. We have to manage as many variables as possible. So the test candidates all must have their lunch at 12.30 and have no more than 1000 Calories. The test will begin by some candidates having their snack at 3 pm. The test begins at 3.15 pm with each of the candidates having the same challenge. Perhaps the challenge is that each candidate has to answer a series of routine maths questions within 30 minutes. Further tests could be carried out later, where the candidates swap roles. If this test falsifies your hypothesis, stop there. If not, you could repeat the study with a different batch of candidates at different times of the year, etc. You will find more information on how to design an experiment in Chapter 7.

Now another argument has broken out over whether students are sleepier, more forgetful and less motivated after lunch than after breakfast. The argument goes around in circles, as some students argue they do better in the afternoon because this is the best time for their body rhythm. So you try to break the deadlock by proposing another hypothesis:

For any lecture or seminar, both the attendance and ability of students to arrive on time for a morning lecture will exceed that for an afternoon lecture/seminar

So for this test you might ask tutors responsible for different classes at the University, starting at the same time, to keep a record of attendance and punctuality in these classes. If it turns out that the performance values are better for afternoon classes, then the hypothesis has been falsified.

You will note that in both these cases we have looked for something you can measure, a parameter that could be considered representative to help form and test the hypothesis.

Testing a hypothesis – what does it mean?

Having proposed your hypothesis, so your experiment (or review of existing data) should be designed purely to test/disprove the hypothesis. There is no need to add any extra bells and whistles beyond what you need to falsify the hypothesis. A classic example is the hypothesis 'all swans are white'. Thus, we only need to find one non-white swan to disprove our hypothesis. We could imagine designing some rapid survey of river and lakes for swans, or perhaps, more economically, check the literature first to see whether black swans have already been observed by others. Falsification is the one absolute thing we are capable of in science. This hypothesis is false because there is a species of black swan.

Getting back to our friend 'Gary' the hard-charging student, I did manage to persuade him to advance a hypothesis, but when his experiments falsified his hypothesis he concluded (wrongly) that the problem lay with his experiment and not his hypothesis. In other words, his hypothesis was so breathtakingly brilliant he was certain the data would inevitably support it. This is where objectivity is vital in a scientist, as we discussed in Chapter 2. In theory, there could be a problem if we prematurely discard a hypothesis due to 'false falsification'. This can occur if an experiment or analytical approach was used that was not suitable to the task and so could only provide misleading results, but during our careers we have not experienced this. If it is practical to study a problem and an appropriate experiment is designed, then we should be able to make progress. We have sympathy for physicists trying to theorise on whether a parallel Universe exists, since they are not in a position to test a suitable hypothesis.

Continuing our previous example, if you cannot find a non-white swan in your river and lake survey, that does not prove that all swans are white. We can never determine the colour of all swans around the world in the past, present or future. The best we can ever say is that the data support my hypothesis. Thus, we should maintain a mind-set that all knowledge is provisional and may be overturned at any moment. It is better to avoid the phrase 'facts' and accept that our knowledge is based on hypotheses not so far falsified.

So I can disprove things? – Absolutely yes!

Can I prove anything? – No, or rather treat a 'proof' with extreme care!

Perhaps we should accept that we can prove things in what might be called 'functional science'. For example, we can describe the world in different measurements and weights. We could prove one car is faster than another due to lower weight or better fuel used. In medicine we can prove the cause of a disease using 'Koch's postulates'. We have proved that the heart is the organ which pumps blood around the body. In physics we can prove the earth moves round the sun and not the other way around. Nevertheless, we would strongly urge that scientists steer clear of making pronouncements that something is 'the truth', that they are dealing in established 'facts', or that something is 'confirmed' or 'validated'. This is potentially dangerous. It discourages challenge and review, which could be damaging to your career if the 'truth' turns out to be false. Our understanding of the world, and where we place our support, is where the strongest evidence lies. You could see this as where a hypothesis has been continually tested and not falsified. Back in the 1970s it was an accepted 'fact' that stomach ulcers were due to stress or spicy food and there was no need to look any further into the matter. So people were treated with medicines that attempted to neutralise the stomach acid. But in 1982 there was astonishment when the hypothesis of stress and spicy food was falsified by Marshall and colleagues, who discovered that the primary cause was a bacterial infection (Marshall et al., 1985). There had been an uncontested assumption or 'fact' that bacteria couldn't live in the acid environment of the stomach. So now, many of these conditions can be treated more successfully with antibiotics.

Getting it wrong

Over the years, as reviewers, we have often been exasperated by reading many grant proposals or papers where the enthusiasm and knowledge of the scientist bubbles over, yet it is still not clear what they want to do, or why they want to do it (despite it being apparently very clear to them). A recent example in a grant proposal had a section of work which described a variety of (rather vague) experiments in a series of paragraphs. A last, almost afterthought, paragraph included what they called a hypothesis, in which the statement made was 'X may play a role in Y' and that they hoped in some way to 'confirm this'. So here they got not one, but three, things wrong! First, their task should have started with the hypothesis rather than ending with it. Second, as we have described earlier, a hypothesis is an unequivocal and clear statement, in which conditional forms such as *may* or *might* do not appear. Third, the proposers cannot 'confirm' their hypothesis, only support it or falsify it. Needless to say, this proposal did not receive wholehearted support!

Many years ago, I used to collaborate with another institute in projects where some complementarity existed and the joint project should, therefore, have been a success. However, the work led to continual frustration. Colleagues from the other institute would propose and carry out a whole series of experiments and, from them, obtain measurements which they piously hoped might do some good, somehow. Finally, when all the work was completed and the money spent, they would end their reports with a ringing pronouncement of their hypothesis. This was their conviction concerning the most likely explanation for their observations. It was there for the reader to take it or leave it; they were the professionals and were convinced they had the explanation and now it was time to move on. Once again, the cart had been put before the horse. The last thing they wanted was a rigorous test of their hypotheses. Needless to say I did not seek to work with them again.

A great skill in scientists is the ability to simplify and clarify the problem so they, and we, may get to the heart of the matter. There is no better way of doing this than proposing a hypothesis. Scientists who offer a hypothesis are doing a favour not only for themselves but also for everyone else involved in their project, from the initial reviewer of their grant proposal to those reviewing their performance and finally for those trying to understand their publications. In later chapters we will review how to write papers, give presentations and write grant proposals. All of them benefit from clear hypotheses! Without it the papers, presentations and proposals will be less valuable than they could be. Regarding your career, learning how to propose a hypothesis will provide you with a magnificent key in your pocket that will enable you to transfer to any scientific discipline. Without a clear hypothesis, science is likely to move forward only slowly and at greater expense than it needs to.

Ultimately, it is better to know something is false rather than to carry on assuming it to be true.

Checklist for Chapter 3: Understanding the hypothesis

1   A hypothesis is a short falsifiable statement, something you can clearly test.

2   Do not confuse it with an aim, objective or theory.

3   Usually, a hypothesis can't be proved, no matter how many positive examples you generate, but it can receive our provisional support.

4   A hypothesis can be disproved/falsified! It only needs one counter-example.

5   Start with as bold and far-reaching a hypothesis as you can. Something that could be really useful if supported by the evidence.

6   As your initial hypothesis is disproved, refine it with alternatives which are more likely to be supported by the evidence.

7   Once you have learned how to form a hypothesis, in theory, you could move into any branch of science. It is the key in your pocket!

Reference

Marshall, B.J., Armstrong, J.A., McGechie, D.B. and Glancy, R.J., 1985. Attempt to fulfill Koch's postulates for pyloric Campylobacter. *Medical Journal of Australia* 142, 436–439.

CHAPTER 4

How do I find my way?

This chapter will consider how you might decide which project and supervisor to choose, as well as how to decide what you should do next and how to develop your science project once you have it.

First stages of selecting projects and supervisors

When you start out on your career you will have very little experience on which to judge either potential supervisors or the research topics on offer. How will you know if they are led by an excellent supervisor and whether the project will be suitable and worthwhile for you? But as all science will require considerable effort from you, you should ideally start by searching for projects that you really like the sound of, that interest you and that you have some preliminary training in. But don't be overly concerned about the level of your background knowledge of a topic, because your enthusiasm and energy will be highly valued by a potential supervisor. They know they can train someone in scientific techniques but they cannot alter your character. There may also be a range of social/human factors that will justifiably affect your choice. These can include the prestige of the institution offering the research position, whether the location is close or far from your family and loved ones, perhaps whether the climate there is conducive, or the nightlife exciting. If the topic area falls roughly within your sphere of interest, we argue that the number 1 issue will be the quality of your supervisor/research manager. To enquire further into both the research project value and the quality of the supervisor/research manager, the modern scientist is now blessed with the many resources of the internet.

Choosing your topic

The ideal moment for a young scientist to get on board is when a subject area is 'taking off'. This is the moment when the importance or concern about an issue is recognised just as the lack of vital information is acknowledged. Searching the internet can reveal the level of interest in the topic in a range of media. If people seem genuinely curious, excited or fearful, then this is a good start. Similarly, a problem may be long recognised but, until recently, no approach or technology has been available to tackle it. When such new approaches become clear, then we have the green light to apply them to the big problem. But this is not the final word. There can be very basic human and national needs which have not yet been resolved by science and that can still make a case for funding. Such a topic might be ways to improve crop production and yields in Africa, or to reduce infections from drinking water in parts of Asia.

Research involves going into the unknown and accepting the associated risk that the project might fail. If you start in a mature field, it is likely that there will be reliable techniques available to guarantee that you will generate data. However, only a few people might be interested in your results because you are following a well-trodden path. On the other hand, an exciting and risky project might leave you struggling to obtain any data. The ideal would be a new and developing field where you can apply robust techniques. Don't feel shy about asking your potential new supervisor these questions!

Choosing your supervisor

After our first degree, most of us assume that all scientists are equally expert and good, and hence any scientist will be capable and motivated to do an excellent job in training and mentoring you. As you may have already gathered, the fact that this is not always the case is one of the primary reasons we are writing this book! We can offer no guarantees but, once again, our friend the internet might help find someone suitable for you.

A prestigious professor might pop up in all sorts of media. They might be spotted giving the keynote speeches at conferences, opening new labs or being seen with government ministers. There is good news and bad news here. The good news is that they are likely to be successful at getting research money, so perhaps could regularly secure funding for you. They will probably also be good at publicising your research to influential people. The bad news is that they may well have little time to help and mentor you. In the worst cases they put their name first on your papers and take advantage of your research to boost their standing. Have a look at their papers on the internet. If they are not the first author on maybe three quarters of them, this would be about right. This suggests they are encouraging their students or researchers to be the lead authors. If they are only an associate author on every paper for the last 10 years, then it could be they no longer have time to be the lead author. Possibly, because of their senior standing, people feel they ought to put him/her on their papers as a way of pleasing them. Either way it may reveal they have neither the time nor the inclination to pursue serious research because they are now more involved in administration, so they are possibly not such a good supervisor for you. If you are clever, you might note the scientists who regularly publish with them. If their contact details are available to you, why not try to tease out what they think of their leader and what the atmosphere in their lab is like? As you are considering applying for a post at that location, these are perfectly reasonable questions to ask. It also will alert the lab in question that you are a serious candidate and no fool.

There is a school of thought that young supervisors are a bad bet. The thinking goes that they will be so desperate to make a name for themselves that they might sell much of your work as their own and perhaps insist on lead authorship for themselves. With much less experience, they may have little clue on best practices in supervision. Alternatively, they have both the time and energy to make you and the project a success. But this is a dangerous generalisation. All one can say is that, with less of a track record to go on, you have much less evidence on which to gauge them as supervisors and inspiring scientists.

Before you commit it is vital you meet a potential supervisor. You can ask them about the project but also how often they meet their students? Also ask to meet some of their existing students to gauge their experience.

Now you are in post, how do you find your direction?

Early days

Most scientists begin their research to tackle an issue or question that another, senior, scientist considers worthwhile or important. To some degree this is reassuring, like a 'duckling following mother'. Your entry into the lab needs to be gentle as there seems to be so much to learn all at once. A good supervisor should appreciate this and start you off with tasks that are straightforward and boost your confidence. Ideally, this will be a set of technical or practical skills that are training you to become more independent. It might be that these tasks directly help another project of the supervisor, so in theory everyone wins from the experience. However, beware of the danger of becoming a resource or pair of hands, used to plug holes in a series of projects which are not your own. If this becomes excessive, you may need to stand up for yourself and point out that your own project is suffering. Let us hope that this won't be the case because these tasks you are doing will be for everyone's mutual benefit. Don't mentally switch off when you are given a task to carry out. Remember, doing science is not an end in itself, it should have some beneficial end-point in mind. Keep asking yourself 'why are we doing this?' Has my work got a hypothesis, is what I am doing leading to a test of a hypothesis? Don't be frightened of asking these questions. If you do that your supervisor will see that you are a serious scientist who wants to improve.

Seizing an opportunity for independence

As you progress in your PhD or further career you will have increasing opportunities to be independent and you must seize these. Even though, based on their boasts at a conference, your competitors appear to have solved everything, you can be reassured that there will never be a shortage of scientific questions that need answering! In fact, if you keep track of those who appear to be hinting at conferences of their amazing breakthroughs, you will see that many of these 'findings' do not make their way into the scientific literature. Your own emotions may sway from alarm to crazy exuberance at the prospect of independence but you are certain to have a voice at the back of your mind asking 'Am I studying the right thing?' One could divide science into small, medium and large questions that need tackling. For example, you could characterise this as questions about:

- the leaf on a tree
- the whole tree itself
- the wood in its entirety.

None of these are mutually exclusive but you would be well advised to recognise that these differences in scale exist.

Choosing a new topic by identifying the big question

When selecting a suitable new topic you must always address the 'so what' test; in other words, why would this research matter to the man or woman in the street? You should force yourself to identify the 'bottom line;' in other words what is it, at its heart, that we really need to know. For example, if our desire is to protect wildlife, the question should be 'what is the greatest threat to wildlife?' It is better to identify and work on the biggest threat and not the most trivial one. Similarly, can you identify the technical bottleneck that prevents us from tackling the issue? For example, with chemistry it might be the lack of a suitable technique that provides sufficient sensitivity to measure a molecule only found at very low concentrations.

One view is that, as all science is hard, whatever topic you tackle, why not focus on something that could have a really big impact and reward? The answer will not come quickly to you. Often, you will need to attend many conferences and meetings and read dozens of papers to get a feel for the subject and the problem areas. But it is best of all to discuss the area/discipline or problem issues with others in your field, who you feel you can talk to and trust. Try to articulate what you think the science problem is and what needs to be done. This can be much harder than you think. Forcing yourself to try to explain something is usually very helpful. The process of attempting to explain through speech seems to assist your brain in getting through the fog of ideas and mass of information that swirls around you. When you are making progress, remember to write things down as soon as possible. All too often, inspiration can be disturbed by the phone, pinging emails and people asking you to come to lunch. Then don't forget to turn your idea or theory into a hypothesis (remember Chapter 3)! But before you charge off, do check that what you propose has not been done already by someone else. It could be that 'great minds think alike' and others have identified the key area you have identified to work on. No shame in that but take a second look – are they missing a trick, could you still take it to a higher level?

The eye of the magpie

Advances in science are rarely the result of inventing something new which has never before been seen by mankind. We would argue that, to make these advances, three elements must come together:

1. Identify a key question.
2. Form a testable hypothesis.
3. Deploy the most appropriate technique.

Finding the most appropriate technique to tackle a problem could mean developing something new but, more often, it is about borrowing a method that has been used elsewhere for a different purpose. Birds like the magpie or jackdaw are famous in the UK for spotting curious or bright objects to steal. This may even mean simply utilising data already generated by others for another purpose to test your hypothesis. Whether you are 'borrowing' or developing something new, a helpful characteristic in a scientist is always having eyes and ears open to learn about different scientific work. This is a major, often unanticipated, benefit of attending those sometimes dull conferences, listening to talks and reading posters.

Don't be put off when others tell you 'it can't be done'. There can be no better motivation for a scientist than hearing this put down! No one has a monopoly on innovative approaches and solutions, not even the most esteemed professors at the most respected institutions. Linus Pauling was one of the most decorated and esteemed chemists of his, or indeed any, generation, with a string of successes in understanding chemical bonds and molecular structures. Yet the Nobel Prize winner for chemistry proposed that DNA was composed of a 3-chain helix, which turned out to be wrong; Watson and Crick's 2-chain helix proved to be the correct answer.

Moving on from your PhD

Once you have submitted your PhD you can be considered, at least temporarily, a world expert in your field. Perhaps you have made some presentations that were very well received at international conferences. At long last you are feeling confident and can think clearly about how to carry forward your PhD topic to yet further success. In other words, why not continue ploughing the same furrow. What an exciting thought, to become the undoubted world grand master of the topic of your PhD! But be careful, is this really the wisest course? You may be on your way to becoming an expert in a topic that the world no longer needs or considers interesting. Alternatively, you are setting yourself up for potentially lifelong competition with your old supervisor, the person who many see as the original 'master', whilst you will be perpetually seen as the apprentice. Can you hear that whispered criticism 'he/she is not original', or 'wasn't able to move on'? So, counter-intuitive as it may seem, our strong advice is to move to a different field following your PhD. This has several advantages:

- It broadens your training and hence expertise.
- Your brain can be excited by the new challenge.
- It exercises all those 'scientific muscles' we have discussed, such as drawing up new hypotheses.
- You start making a whole new circle of contacts.

This experience will increase your employability. All potential employers will rate those who can demonstrate they can adapt to new environments and learn new skills.

It may take a lot of courage to do it but that courage will nearly always be repaid many times over.

Checklist for Chapter 4: How do I find my way?

1   Start with areas of science you really like, that interest you.

2   Maybe look for topics which are just starting to receive a lot of interest.

3   Select the right supervisor for you.

4   Seize opportunities for independence when you can.

5   Look for big or worthwhile questions to study, will they pass the 'so what' test?

6   Keep an eye out for applying techniques from other disciplines/areas to your problem.

7   Consider changing topics as your career develops, to widen your skills and knowledge.

CHAPTER 5

Integrity in science

Science is arguably society's greatest calling and humanity's greatest asset. It is an immense privilege to be a scientist and add to humankind's knowledge. Human nature may not have changed much over the last few thousand years, but human knowledge certainly has. Science builds knowledge whilst eliminating falsehoods; that is what differentiates science from many, if not all, other disciplines. The information provided by science is universal and hence relevant to all of society. In order to provide knowledge that can be relied upon, science depends absolutely on the integrity of scientists. Nothing is more important in science than integrity. Acting with integrity means being not only completely honest but also completely transparent. By transparent we mean that nothing must be intentionally forgotten, hidden or avoided. For example, if you make a small mistake while conducting an experiment – and we all do – record the details of that mistake and tell your supervisor and anyone else who needs to know about it. Do not conveniently omit to mention it. Good supervisors will welcome your honesty (see Chapter 15) and think more highly of you because you were open and honest with them. However, scientists are only human, and hence they do not always conduct and/ or present their research to the highest standards. In fact, there is strong, and rapidly increasing, evidence that a significant proportion of published research in many areas of science and social science is poor and that the results may not be reproducible. It has even been claimed that '...most published research findings are false' (Ioannidis, 2005). That may seem a harsh judgement and it does not imply that all published research is full of major errors, but it does suggest that the quality of published research could, and should, improve, and that many of the current problems leading to poor research are a consequence of low ethical standards leading to a lack of integrity. Certainly, what appears to be a rapidly increasing rate of retraction and correction of papers published in the premier scientific journals may indicate that there is a significant problem with reproducibility. Not all of the irreproducibility is a consequence of lack of integrity – it may, for example, be due to chance – but a significant proportion appears to be. Lack of integrity can range in degree from an unconscious bias or a less-than-ideal experimental design through to outright fabrication of results. Even the total number of 'minor' mistakes in a paper can be indicative of major problems with that paper (see Nowbar et al., 2014). In the paragraphs below we discuss some of the most important issues concerned with maintaining integrity in science.

The pressure to deliver good results

Most, possibly all, scientists feel under pressure to produce 'good' results. The more novel and interesting their results, the more attention those results will receive and the more kudos the scientist will receive. That additional kudos can, in turn, lead to an enhanced reputation, tenure, promotion, more success in raising research funds, and job opportunities. The rewards are high!

Similarly, in many cultures it is expected that juniors (in this case junior scientists) should please their superiors. They can react to this cultural norm by tweaking their results so that these support the hypothesis of the supervisor. Sometimes the junior scientist, aware of the prevailing consensus, will do the utmost to ensure that results fit in with the agreed view on the topic. In one peculiar example, I examined a PhD thesis where the candidate demonstrated that a particular pesticide, contrary to their own and perhaps their supervisor's expectations, had no harmful impacts on non-target species. Despite the evidence, the student concluded that the pesticide was dangerous to wildlife and that more research was needed, wrongly believing this message would please the examiner!

All scientists are aware of these pressures, yet they must resist them and maintain a high level of personal objectivity and integrity at all times. Succumbing to these pressures can lead to publishing results that other scientists cannot repeat, which is not something you want to do (see Chapter 7: The reproducibility crisis). Sometimes, in extreme cases, this pressure from supervisors and advisors can be explicit; there can be a feeling that only 'good' results are acceptable to senior scientists and hence there is a strong temptation to manipulate results in various ways to make them look better. Resist this temptation! Young scientists must always present the whole story to their seniors, even if that story is not what everyone hoped that it would be. Do not distort your results before presenting them to others. If you feel under pressure from a superior to assist in misleading the wider scientific community, then you know that you are not in the right place to train as a scientist: as soon as you can, move to work elsewhere, where there are people you respect.

Designing experiments and analysing results

The design of experiments is crucial, as is the analysis of results. Although many young scientists now receive at least some training in statistics as an aid to analysis of their results, very few receive any training or guidance in experimental design, unless it comes informally from their supervisors. Yet poorly designed experiments can easily lead to integrity being severely compromised. For example, if the appropriate controls are not included – and you would be surprised how often they are not – then nothing can be concluded from that experiment and to try to conclude something is not only poor science, it demonstrates a lack of integrity. It can be very helpful to keep your hypothesis in the forefront of your mind (see Chapter 3) when designing experiments, so that your experiments can provide results that test that hypothesis. Remember that, before conducting an experiment, you need to be competent in all the techniques that will be utilised to complete the experiment, which includes fully testing those techniques beforehand, so that they produce consistent, accurate results. Do not be in a hurry to produce results: if you conduct a well-designed experiment that utilises validated techniques, it is highly likely that you will obtain good (i.e. reproducible) results that do not need to be manipulated in any way before you show them to other scientists. That way you avoid the temptation to 'improve' the results by any unethical means. 'Tidying up' data is unacceptable.

Appropriate analysis of results is very important. Superficial analysis of results is not only lazy, it is unethical. Depending on what type of experiment you want to conduct, it can very often be wise – some would even say mandatory – to seek the help of a statistician when designing the experiment. Seeking his or her help after doing the experiment is poor practice; by then it is too late to repair the damage! Do not rely on statistical software packages to provide analysis of your results if you do not understand the statistical tests utilised. The best way to maximise the integrity of your research is to put a lot of time into thinking about what you intend to do and how you will do it before you go into the laboratory and begin your research. Always seek help and advice from others you respect.

A few examples:

A. Imagine that you have conducted a well-designed experiment utilising fully validated techniques to obtain a set of numerical data, yet it is not quite as good as you hoped it would be: it is spoilt by just one value in one experimental group, which is very different from all the other values in that group. It is an outlier. What should you do? There will be a strong temptation to simply omit that single data point and not admit to anyone, particularly your seniors, that it exists. Do not do so. Be open and honest about that outlier. Tell your supervisor about it and discuss it with a statistician. Decide how to handle it and admit to it in any subsequent publication. It may even be telling you something very important.

B. Imagine you have conducted three experiments. Two experiments have generated the results you expected and hoped for, whilst the third surprised you by appearing to produce the opposite result. Perhaps you might think to yourself there was some mistake in the third experiment, so it would be better to deny it ever existed. After all, a journal may accept your work based on only the two experiments. But to do this would be deceitful.

C. You may be half way through an experiment, or even have completed it, when you realise that you have made a mistake. Perhaps you made up a solution incorrectly, forgot to include a standard when conducting the analysis of the samples, or mixed up some samples. Do not try to 'muddle through' in the hope that nobody will notice, and all will be alright. Admit your mistake. Good supervisors will not blame you for it; instead, they will praise you for your honesty, which will strengthen your relationship with them and in turn lead to better science from both of you.

D. Imagine you are conducting an experiment involving a time course: you are sampling or measuring after 1, 4, 12, 24 and 48 hours. Or should be! But you overslept as a consequence of being very tired due to the demands of the experiment, and you miss the 12 hour sample. Instead, you take it after 15 hours. Do not record the data as though they were collected after 12 hours.

It is inevitable that not everything will go as planned and hoped all the time. Good scientists realise this, and are open and honest about the unexpected problems that occur.

The issue of bias

Any psychologist will tell you that humans are remarkably good at self-deception. We may think that we are unbiased (but know that everybody else is!) but we are wrong. We all have our biases, both conscious and unconscious. The former is usually a consequence of us already knowing the result we want, so we interpret our results in a way that supports our conviction. It should be the data and facts that guide us, not our beliefs and presumptions (our biases). This jumping to unwarranted conclusions, rather than considering alternative explanations, is surprisingly difficult to avoid and is where a statistician can be extremely helpful in steering you to an unbiased opinion based on appropriate analysis of your results. Unconscious and unintended biases arise when factors other than those under study influenced the results. These factors can be both known and unknown. If known, then their influence can often be minimised through good experimental design but, if unknown, they cannot be controlled for. The way to deal with bias is to be open and honest about it; do not pretend that it is not there.

Some fields of science can be extremely contentious, with scientists possessing radically different opinions. Climate change is an example. Any scientific field which has political relevance is quite likely to be contentious. Even if you have your own opinions, as is very likely, try your best to conduct, analyse and report your research in as unbiased a manner as possible.

Presenting and publishing your research

It is surprisingly easy for integrity to take a back seat when a scientist prepares a presentation for a conference or writes a paper. There are a number of reasons for this. A major one is that the scientist will want to give the most exciting talk he or she can, or want to maximise the chances of their paper being accepted for publication by the journal they submit it to. The more prestigious that journal is, the greater the temptation to compromise your integrity and act unethically. Another is that the time restriction of a presentation, or the word limit imposed on your paper by a journal, may be used as an excuse to omit information. Yet despite these pressures, the objective must always be to present as balanced and representative a picture as possible of the results of your research. Do not ignore relevant literature, either because it reached different conclusions to those you reached or because it comes from your competitors. Anyone who has reviewed papers for journals soon becomes aware that this type of unethical behaviour is very common. Try to resist the temptation to hype your results – meaning exaggerate their significance, perhaps by making them sound more original and exciting than they are – in the hope that doing so will increase their impact. Scientists are very good at spotting hype and will think less of your paper or presentation, and you, for acting in this manner.

The pressure to publish can easily lead to premature publication of results. Do not publish your results until you have completed a study and have confidence in the results; perhaps through replication of the experiment prior to publication. Or, if you do present some results from your research before the study has been completed, be honest and willing to describe your results as preliminary. Many sound and influential scientific papers have the words 'Preliminary evidence' in their titles! It can be refreshing for the reader when the paper has a section where the authors discuss the limitations in their own study.

Because the number (rather than the quality) of papers a scientist has published can be considered by some as a reflection of the quality of that scientist, there is also a temptation to publish the results of a single study in more papers than are necessary and justified: this practice is called 'salami science' and is unethical. This is not to be confused with our advice (Chapter 10) to write papers preferably with a single, clear message (supported by strong evidence). Several papers that essentially give the same message, with minor variations, will be much less highly regarded. An uninformed assessor of your publication record might conclude 'this person is a productive scientist' but knowledgeable people, such as other scientists in your field, whom you will be hoping to impress, will easily identify what you have been guilty of and will probably think less of you because of it.

It is quite possible that, after publishing an article, you discover that it contains an error: you might have realised this yourself, or another scientist may have made you aware of it. Correct that error. Not to do so demonstrates a lack of integrity. You do not want to be responsible for intentionally misleading other scientists. Journals are becoming progressively more willing to publish corrections of errors, and should encourage and help you correct any errors in your published work (Allison et al., 2016). Correcting errors does not diminish your reputation amongst other scientists; instead, it is likely to enhance your reputation in the eyes of those who want science to reflect the truth as far as possible.

Peer review of the research of others

Quite early on in your career as a scientist you may be asked to review papers that have been submitted to journals for possible publication. When you review papers (and grant applications), do not impose much higher standards on your fellow scientists than you apply to your own research. Doing so is both unfair and unethical. Aim to provide objective and constructive criticism if justified. Focus exclusively on reviewing the science, and try to avoid letting any personal jealousy, previous disagreements with the authors, the fact that the paper comes from your competitors, or the fact that you have been 'scooped', influence your review. This is much easier said than done, of course. And when you receive reviews of your papers, respond positively and constructively to them: you will be surprised how often 'negative' and 'unhelpful' comments can improve your paper!

Conflicts of interest

It is becoming increasingly apparent that conflicts of interest are both very common and that they lead to biased reporting of scientific research. That is why most scientific journals now request that authors provide a 'Conflicts of Interest' statement. Many conflicts of interest arise as a consequence of the funding source of the research. For example, if a study is funded by industry, there can be pressure, both conscious and unconscious, to report results favourable to that industry. But conflicts of interest can originate in a wide range of circumstances, not just the source of funding, so think hard about any potential conflicts of interest you may have and be open and honest about them. These conflicts, or competing interests, in science include possible financial gain and maintaining professional relationships. These can compromise the upholding of ethical scientific practices such as reporting data accurately and completely, interpreting your data appropriately, and acknowledging any limitations of your research.

Training in ethics and integrity

By now, having read this chapter, you will realise that there are many different aspects to scientific integrity and that maintaining high ethical standards, although obviously desirable, may not be as easy as it sounds. The recent realisation that integrity is not as embedded within the research community as it could, and should, be has led many universities and research organisations, as well as some scientific societies, to develop codes of ethics (or ethical behaviour), and also to offer formal training in ethics. If such support is available in your organisation, take full advantage of it: sometimes it is now mandatory for young scientists to take these courses. All research centres should build a culture and infrastructure that strongly encourages integrity.

A very brief summary

Nothing is more important for science, and the faith society places in science, than that scientists have integrity and act ethically. So it is very important to instil in young scientists strong ethical principles, so that they apply these to all aspects of their research, from planning experiments right through to publication of their results. Other scientists (often your primary audience) will be able to tell good science from bad, just as you can. If you want to make a reputation for yourself as a good scientist, it is vital to ensure that your research is done to the highest possible standards achievable in your circumstances. Aim for nothing less. You will usually know when your decisions or actions are compromising your integrity: avoid getting into that situation.

Checklist for Chapter 5: Integrity in science

1   Without honesty science will fail.

2   Design your experiments as thoroughly as possible, taking particular care over the controls.

3   Resist pressure to overlook or ignore awkward data.

4   Be the first to admit errors or mistakes.

5   Don't see things which aren't there because of the 'eye of faith'.

6   Honesty and transparency will highlight your qualities as a good scientist.

References

Allison, D.B., Brown, A.W., George, B.J. and Kaiser, K.A. 2016. A tragedy of errors. *Nature* 530, 27–29.

Ioannidis, J.P.A. 2005. Why most published research findings are false. *PLoS Medicine* 2, e124.

Nowbar, A.N., Mielewczik, M., Karavassilis, M., Dehbi, H-M., Shun-Shin, M.J., Jones, S., Howard, J.P., Cole, G.D. and Francis, D.P. 2014. Discrepancies in autologous bone marrow stem cell trials and enhancement of ejection fraction (DAMASCENE): weighted regression and meta-analysis. *BMJ* 348, g2688.

CHAPTER 6

Lack of confidence and the embarrassment factor

People vary greatly in the amount of confidence they have in their abilities. Some people seem to be extremely confident (but are they?), whereas others appear to lack confidence in almost everything. Most people are probably somewhere between these two extremes: they are confident in some things but not others. As people progress through life, the degree of confidence they have can change; often they become more confident, especially in things they have done previously and hence learnt that they can do, often more easily than they thought they would be able to.

Knowing little when you start

It is inevitable that you will know relatively little about the topic you have chosen to study when you begin your scientific career. You will probably have relatively little theoretical knowledge and also little, if any, practical experience relevant to the project you are about to embark on. This is normal: do not worry! At this stage of your scientific career by far the wisest course of action is to admit your lack of experience – and hence quite probably also confidence – and need for assistance straight away. Honesty will always be well received, and a supervisor can then consider the appropriate way to help you gain the competence you will need. Try not to be embarrassed by your lack of knowledge and experience.

Besides your supervisor(s), you can learn a lot from others, particularly other young scientists and technicians. It is quite possible that there is a wealth of talent close by! It is very likely that you are not alone: you may well, for example, have your desk in a room where other young researchers are based. There may be research assistants and/or research technicians available to provide advice; the latter may have the training of young scientists as part of their job. More experienced scientists, at different levels, hopefully with friendly natures, can also help and advise you. For them it is flattering to be asked to give advice or share experience. Seeking help and advice is not a failure; it is wise.

Nobody knows everything

At the beginning you may think that you know nothing whereas many of the scientists around you appear to know everything. Neither assumption is true. You probably know more than you realise and nobody, however intelligent and knowledgeable they seem, knows everything. Both of the authors of this book have learnt a great deal from much younger, less experienced scientists than themselves. In fact, it is often the inexperienced scientist who has the most interesting, and novel, ideas, because her or his mind is not full of what others have already said or done. Hence, if you are not very confident about your knowledge and abilities, nevertheless try to overcome your embarrassment and contribute to discussions: you will not be the only one to benefit if you do so.

The over confident scientist

Some scientists, both inexperienced and experienced, can be over confident. Whilst having strong confidence in your ideas can help drive you forward, this can go too far and become dangerous for science. The danger is there because the views of an over-confident person may be wrong and that person may be unwilling to consider opinions different from their own. Their inflexibility makes it difficult for over-confident scientists to consider new information when it becomes available, especially if that information contradicts their own opinions. It is not unknown for apparently highly intelligent people to stick with their personal convictions on an issue despite knowing of overwhelming evidence to the contrary; you need only to think of deniers of climate change. One hallmark of a good scientist is the ability to change an opinion when evidence emerges suggesting that opinion is not correct. Scientists who are not particularly confident (or who remain resolutely objective) find this easier to do than do over-confident scientists.

Confidence issues

Confidence is probably an innate characteristic, meaning that it is deeply established in people's basic characters. This means that it can be very difficult to acquire. Nevertheless, there are ways of preparing the mind to take small steps toward feelingly sufficiently confident to get our work the publicity it deserves:

- Motivation can build by viewing the efforts of others to explain or promote their work. It won't escape your notice that many folk are poor at explaining their work or, in some cases, they promote its value beyond its true worth. This should tell you two things: (a) the current standard of many scientists as confident advocates of their work is not that high and (b) some poorly prepared and poorly conceived work receives greater attention than yours; why should it?

- Seek support at home and find enjoyable distractions. Various issues can reduce your performance and confidence. No two people are alike in how they manage their anxieties/fears/concerns. For some, yoga and meditation can help. For others it is the gentle reassurance of parents, partners and friends. Perhaps even taking up gardening or having a pet could help build you back up. We certainly want our scientists to have full lives outside the laboratory. We would not recommend resorting to alcohol or drugs; as a scientist you will know the potential pitfalls! There are a number of excellent books which promote positive thinking about yourself and they could help too.

- There are opportunities for taking small but confidence-building steps in science. These include standing by a poster or giving a talk to colleagues in your department. These will help to draw you out as a scientist who can explain and defend your work in public.

Even if you don't feel ready yet to leap onto a stage or write a major research grant application, these days nearly all research is done in teams. These teams most often comprise an independent scientist and several junior scientists, all contributing to one goal. Those who are initially less confident can still have a major role to play in these teams by supporting the research agenda of the independent scientist. It is extremely important to realise that confidence and competence are completely different characteristics. Both authors of this book have worked for many years with young, and not-so-young, scientists who are not very confident but who are extremely competent. We, and most independent scientists, rely on these scientists to do 'our' research; hence, we value their contributions greatly.

The 'never satisfied' scientist

A lack of confidence can sometimes manifest itself as an inability to 'let go' of a piece of research. People who struggle to do this are perfectionists: to them nothing is ever as good as it could and should be, and hence they constantly attempt to improve a piece of research that most other scientists would consider finished, and hence ready for publication, long ago. They never reach the 'enough-is-enough' stage, when a piece of research is considered complete enough to merit publication. These people are, in fact, usually extremely competent, but nevertheless do not have the confidence to cease doing yet more experiments or making endless minor modifications to a paper they hope to publish. It is important to realise that no research topic is ever 'finished' and everything completely established; there is always more that could be done. So you need to learn to decide when 'enough-is-enough', when it is time to stop doing yet more experiments and start writing up and submitting work to the science community. Ask other scientists you know for their opinion on whether the quantity and quality of data you have is now sufficient to weave together into a good story for a paper.

Restricting your research due to lack of confidence

Scientists lacking confidence tend to focus all their attention on the one area of research that they are active in; they do not have the confidence to broaden their research, let alone change fields. Yet reasonably confident scientists can probably contribute to other areas of research besides their own, and these contributions may even be novel and hence very worthwhile. This is because their minds will not be full to overflowing with all the detailed knowledge that is part of any field of science. They may bring new ideas, new approaches, and novel concepts to areas of research other than their own. So, once you have gained confidence in your abilities as a scientist, do not be afraid to contribute to other areas of science when suitable opportunities arise. Remember, being able to construct a testable hypothesis (Chapter 3) enables you to contribute to any scientific field.

The crucial role of a good supervisor(s)

It would be very difficult to overstate the role of good supervision in helping young scientists build their confidence. Supervisors should not expect too much initially.

They should realise that you are likely to know relatively little about the topic you are about to study and also that you will have little, if any, practical experience of the approaches and techniques you will be utilising during your research. After all, they were in your position when they began their research careers (although some might have difficulty recalling this). Supervisors should be supporting and encouraging their research students, so that the latter grow steadily in confidence. This includes listening to them (see Chapter 15)! Supervision should not involve telling research students 'Do A, B and C by next week, then come and see me again'. Such a strategy, if it can be called a strategy, will demoralise a student, not encourage them. It takes time to grow as a scientist (see below), so be prepared to gain in confidence only slowly but steadily. Be honest with your supervisor(s); if you do not know, or understand, something, then say so. The better your supervisor(s) understands you, and what you know and do not know, then the easier it is for them to help you.

Learn to respond positively to constructive criticism

Nobody is perfect, so it is inevitable that things will not always go as planned; that may be your fault, the fault of others, or nobody's fault (equipment may break down, for example, or not function correctly). In such situations, you may receive constructive criticism from your supervisor, if he or she considers it deserved. Try not to be too deflated by such criticism. Accept criticism if it is justified and respond positively to it. You will be able to learn from it, which will help build your confidence and hence support your development as a scientist. As you will discover once you submit a paper to a journal, hoping that it will be accepted for publication, anonymous reviewers can be very harsh critics of the work of others (i.e. your research!), so the earlier in your career you build the confidence needed to cope with criticism, the faster you will develop as a scientist (see Chapter 12: How to cope with rejection).

Gaining confidence through presenting and publishing the results of your research

One of the best ways to gain confidence in your scientific abilities is to present the results of your research at a conference, either as a poster or as an oral presentation, or to publish them in a scientific journal. Giving a presentation, either in your own organisation or at a conference, can help a young scientist gain in confidence, even if the thought of doing so terrifies you! If you are to become an independent scientist, it is necessary to overcome this understandable fear (although in our experience it never goes away completely) and present your work to others. If giving an oral presentation seems too daunting, you might instead choose to present your research on a poster, which many young scientists find less stressful (see Chapter 9: Giving a presentation or a poster, for practical tips on how best to present your work to others). It may seem intuitively unlikely but experience will demonstrate that by presenting your research you discover two important things: that other scientists are interested in your work and that you can find ways to answer any questions they ask. This will help greatly in building your confidence. Publishing your first scientific paper also helps a lot to build your confidence; you will be inspired when you see your research and your name in print and available to everyone and hopefully enthused to write the next one (but only once you have something further to say, of course). The 'art' of writing a scientific paper is discussed in Chapter 10: Writing a (good) paper.

Becoming confident takes time

Do not expect to become a confident young scientist a month after beginning your PhD! It takes years, not months or weeks, to become a confident scientist. But if you persevere and learn from those around you, you will achieve it. This confidence could come from allowing yourself to enjoy the thrill of being at the cutting edge of science or recognising that most, if not of all, of humanity currently do not know the answer to the problem you are studying, that your footsteps could be the first ones in the snow of a winter's morning. If you can express to others the excitement of your discoveries, then you are well on your way to becoming an independent scientist.

Checklist for Chapter 6: Lack of confidence and the embarrassment factor

1   All scientists start by knowing little.

2   Even the best scientists don't know everything.

3   Over-confident scientists can be a menace.

4   Examples of under confidence include being unwilling to present your work, not letting go of datasets, and restricting your research, lifelong, to a small area.

5   Be open to your supervisor and/or mentor about support you might need with confidence.

6   Science requires everyone to accept a need for constant improvement, so don't take criticism too hard.

7   Plan baby steps with tasks and presentations that will gradually increase your confidence.

8   Confidence takes time, so stick with it!

CHAPTER 7

The basics of doing an experiment

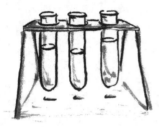
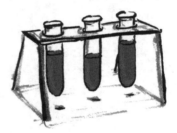

Not all scientific research is experimental, but much of it is. Some may be observational rather than experimental. For example, if your research is concerned with climate change, it might involve observing whether or not glaciers are retreating, or if it is concerned with biodiversity loss, it might involve assessing the size of animal populations and determining how these have changed over time. But many scientists gain their scientific training by doing experiments. Because experiments are so important to your training and also to your development as a scientist, you would think that, early on in your career, you would be taught experimental design; yet, surprisingly, this rarely occurs. Most supervisors and research organisations do not provide formal training in experimental design, yet it is central in many of our attempts to advance scientific knowledge. A well-designed experiment will provide results that can be interpreted with confidence (even if they are not the results you hoped for, or expected), whereas a poorly-designed experiment cannot provide useful results that are likely to be repeatable. Both authors of this book have reviewed hundreds of papers that have been submitted to journals in the hope that they will be accepted for publication, but have not been able to suggest acceptance of those papers because of poor experimental design leading to results we have little or no confidence in. Below we discuss the key components of a well-designed experiment. Not all of them will apply to all experiments, but an awareness of factors that contribute to a good experiment will prove invaluable to you, whatever type of research you are doing.

Start with a hypothesis

It is often best to start with a hypothesis (see Chapter 3) and design an experiment to test that hypothesis. But whereas your project may have one main, quite ambitious, hypothesis (e.g. chemical X caused the loss of species Y in the environment), less ambitious, more specific hypotheses are usually the best starting point for individual experiments (e.g. chemical X suppresses concentrations of hormone A in species Y). It is much better to test a hypothesis, however simple that hypothesis might seem, than it is to be thinking something like 'I wonder if chemical X has an effect on species Y', and then conduct an experiment to probe the issue. The clearer your thinking is at the outset, the better the experiment you are likely to design and conduct, and the better the results will be. So think clearly before designing and conducting an experiment. The time spent thinking and discussing your design with other scientists, including your supervisor, will be repaid many times over.

The role of a preliminary experiment

It is quite possible that, especially at the beginning of your project, you will know little or nothing about the likely outcome of your proposed experiments. For example, you may not know if chemical X has any effect on species Y, or if an environmental pollutant is adversely affecting the wildlife exposed to it. In such cases there can be a very strong argument for conducting a preliminary experiment, the aim being to obtain from a small-scale experiment information that will be used to aid the design of a larger, more definitive experiment.

Conducting a preliminary experiment can have other advantages. One is that it allows you to practise all the steps involved in your experimental research and thus you will be more proficient when you do conduct a larger experiment, and you will have a much better understanding of what is practical and what is too ambitious. Another is that unexpected occurrences that compromise the objective of the experiment may be discovered. These can then be taken into account in the design of subsequent, larger experiments. It is better to learn of problems from small, preliminary experiments – which will use less resources – than it is to encounter them when you are in the middle of your main experiment. We strongly recommend that you consider conducting a preliminary experiment before you conduct what you hope will be a definitive experiment. If you obtain basically the same result in the preliminary experiment and in the subsequent main experiment, that will give you a lot of confidence that your results are robust; that is, they are repeatable.

Keep it simple

Whereas it is probably impossible to conduct an experiment that is 'too simple', it is certainly possible to conduct an experiment that is too complex. By far the most common mistake made by young scientists is to attempt to do far too much in a single experiment. Do not try to answer ten questions in one experiment. Do not try to answer even two! Try not to be over-confident because your preliminary experiment went well. Keep your experiments as simple as possible. As I learnt from my own – very good – PhD supervisor 'simplify, simplify, simplify'. These are the words I have said, often many times, to every one of my own research students. They are usually initially reluctant to accept them, but they soon realise that those words are crucial to the design of any experiment. You may want to know how several different factors (they could be different chemicals, or different environmental conditions, for example) affect whatever you are studying, but try not to fall into the trap of designing an experiment that investigates the effects of all these factors simultaneously, through a multi-factorial design. You may think that such an experiment will answer all of your questions in one go but I am very confident that it will not. Resist the understandable temptation to have many different variables under investigation in one experiment. A series of relatively simple – and smaller – experiments is much more likely to provide reliable, repeatable results than one or two larger, complex experiments. Further, the results from your first 'simple' experiment can be used to aid the design of your second 'simple' experiment, and the results from that experiment used to aid the design of the next experiment, etc., etc. Thus you steadily build up a set of robust results that you can have confidence in, with each additional set of results supporting and complementing the previous ones.

Include enough controls

Controls are crucial to any experiment. You need to have enough controls, as well as having appropriate controls (see below). The controls are as important as all the other treatments (e.g. different doses of a test chemical, different degrees of an environmental factor, etc.), yet it is very common to come across experiments in the literature where there were perhaps only six control animals, or wells in a petri-dish, but well over 50, or even 100, treated animals or samples. If those six controls are not representative, you will be comparing the treated animals or samples with an unrepresentative set of controls, which could easily lead to you reaching unreliable conclusions. You need enough controls to provide you with an estimate of the variability within the (untreated) controls of the endpoint you are studying. For example, if you are assessing the possible effect of chemical X on the concentration of hormone Y, you need to know how variable the concentration of hormone Y is in untreated animals or samples. This is what the controls will tell you. You may decide to include more than one set of controls in an experiment. An example of the crucial role played by controls in any experiment can be found in Owen et al (2010).

We need to be confident that the transformation or change in the subject we are studying is due to the driver/variable that the experiment is trying to test. A concern is that the exciting results obtained are due to what is known as an artefact of the experiment. In this case we are not revealing a natural phenomenon but an accident of the design or materials of the experiment. Thus, the range of controls we select must be based on isolating the widest range of components used in our experiment. For example, we must be sure that the transformation was not down to something peculiar in our glassware, bathing solution or food we might have given to a test animal? So we should add controls to enable us to examine the impact of these components on the experiment in the absence of the driver/variable we are hoping to study.

The usual controls are known as negative controls; for example, the untreated animals or cells. These controls will tell you what is normal. But it is often possible to include positive controls in an experiment; these controls are treated with something that is well recognised as affecting the endpoint you are studying. If the expected effect is produced by the positive control, then this provides reassurance that the experiment is sound. For example, if your experiment was designed to determine if a particular chemical had estrogenic (feminising) properties, then including a set of samples (animals or cells) that were exposed to a known estrogenic chemical, such as estradiol, would very significantly improve the design of the experiment. If estradiol caused the expected effect, then you could have confidence that the experiment would detect any estrogenic activity in the chemical of interest. Equally, if the chemical of interest did not cause an effect, then you could be confident that it did not possess estrogenic activity; at least not at the concentrations you tested. Both negative (no response) and positive (a response of known magnitude) controls should be included in all analytical techniques. Unless these samples are included, and give the expected results, you cannot be sure that your analytical techniques themselves are giving you reliable data.

Fieldwork can often present major challenges when it comes to including appropriate controls. Imagine, for example, that you want to know if a particular chemical known to be present in the environment was adversely affecting a particular group of animals (e.g. fish, birds). Ideally you would include in your study a number of sites where the chemical was known to be present and a number where it was known to be absent. Then you could compare the exposed animals to the unexposed ones. But many chemicals are present everywhere and hence there are no unexposed sites that you can use as your controls. Often, the best option in such situations is to include in your fieldwork a number of sites with different degrees of contamination, ideally ranging from slightly contaminated to heavily contaminated. If your endpoint of interest is found to correlate with the degree of contamination, then you have evidence supporting the hypothesis that your chemical of interest is causing the effect. But always keep in the forefront of your mind that association is not necessarily causation. There is rarely, if ever, a perfectly designed fieldwork experiment. Good scientists realise this, yet seek to design the best experiment (or survey) possible, always remaining aware of the limitations of their fieldwork.

The importance of sample size or n=?

Science is about persuading not only yourself and your best friend but also the wider, sceptical science community. Has the effect you have observed really been a genuine change (and so repeatable) or was it due to an extraordinary chance? In statistics, this danger might be called a false positive or 'Type I error'. But a more common problem is that the effect is small and you overlook it, which is called a false negative or 'Type II error'. Statistical power is the likelihood that a study will detect an effect, assuming it exists. This is improved for small effects by having a large sample size. Thus, we should attempt to have as many subjects, samples and replicates in our study or experiment as possible, to reduce the chances of being misled by a random event or overlooking a small but important effect. Statisticians can help you by advising how many samples/subjects you would need in your study to observe an effect of a certain magnitude with a selected level of significance. The answer may be many more are needed than you expected! This is a topic where the general public often misunderstands science issues. You only start persuading your fellow scientists when you can show a great many observations which repeat the same effect. There is a nice statement which goes along the lines of 'extraordinary assertions require extraordinary levels of evidence'. A cautionary tale comes from the infamous MMR and autism story where a doctor in UK asserted that there was a link between autism in childhood and a vaccine called MMR. He based his assertion on a handful of examples (eight). Many in the public and media found this compelling, despite the many thousands of individuals studied around the world where no link could be shown (DeStefano, 2007).

Seek statistical advice early

Far too many experiments are performed without the person conducting the experiment having any idea about how to analyse the data collected. It is very common for inexperienced (and sometimes experienced) scientists to conduct an experiment, collect the data, and only then think about how to analyse these data. That is poor science. You should think carefully about how you intend to analyse your data before conducting your experiment. You do not want to seek the advice of a statistician after you have done your experiment, only to be told 'I would not have done the experiment that way'. Talk to a statistician, or to someone who is statistically competent, when you are designing your experiment. Obtaining his or her input at the beginning of your experiment will ensure that your data can be analysed appropriately and robustly at the end of the experiment. For example, a statistician will advise you on the likely statistical power of your proposed experiment; put simply, what are the chances that your experiment could produce statistically significant results? Sample size can be a key factor here and a friendly statistician will be able to help you decide on an adequate sample size so that robust results can be obtained. Far too many reported experiments are under-powered, and hence have little or no chance of providing robust, repeatable results.

The dose–response relationship

We work in very different branches of science but in many fields we presume that the greater the input, the greater the response. This is encapsulated in ecotoxicity in the dose–response relationship. Experiments often provide particularly convincing results if it is shown that the results are dose (or concentration)-dependent; that is, the higher the dose, the greater the effect. Not all experiments enable the dose–response to be assessed, but many do. If you are able to test more than one dose or concentration of the factor under investigation in your experiment, then do so, but do not make the experiment too large, and hence unmanageable, while doing so. If you test a number of doses or concentrations, think carefully about what doses or concentrations to test. It may be best if they are evenly spaced apart and not too similar to each other. For example, testing doses of 1, 10 and 100 is usually much better than testing 1, 5 and 100, and definitely much better than testing 1, 2 and 100 – although there are plenty of examples of the latter in the scientific literature. This is another aspect of experimental design that your friendly statistician will be able to help you with.

Bias

As we start our experiment we often entertain great hopes that it will succeed and support our cherished hypothesis. Inadvertently, we then have a bias that tends to make us see what we want to see (and ignore the inconvenient). This can start right at the beginning with not including appropriate and sufficient controls (since we are convinced of what will happen in the experiment). We can influence ourselves and others doing the analysis when our test tubes are clearly labelled as controls or treatments. We are sending a message to the analyst about what we are expecting to get back. The analyst then may wish to please us by down-playing, or ignoring, results which seem inconsistent, so that you get 'the right results'. The medical world is painfully aware of this danger and so insists on blind or double blind trials in the tests of new drugs. In the ultimate double blind trial, the clinician, the patient and the analyst do not know the code telling them which is the drug and which the control (placebo). Where an end-point could be considered ambiguous, or requires judgement, such as in histology (looking down a microscope at a tissue specimen), it is wise to get a third party to corroborate the observations made by your team. There is a lovely phrase I heard many years ago from a senior scientist, something he called 'the eye of faith'. Here scientists persuade themselves and others that a rather inconclusive set of data still somehow adheres to an expected pattern. Beware!

The reproducibility crisis

In the last few years it has become apparent that a significant proportion of published scientific research contains results that other scientists cannot reproduce. There are many reasons why this is the case, with a major one being the pressure on scientists to publish results before they have assessed their repeatability in order to quickly gain kudos and improve their chances of obtaining a better job, promotion in their existing job, or research funding (see Chapter 5: The pressure to deliver good results). In many countries a PhD cannot be obtained unless the candidate has published a certain number of papers. Similarly, the number of papers a post-doctoral scientist has published can be a major factor in determining whether or not they are able to secure a permanent scientific position. The pressure to publish papers is high and science has become very competitive. Perhaps it is therefore not surprising that these days a lot of science is published without the authors having demonstrated that the results are repeatable. Many papers contain results from only one experiment and hence it is impossible to know if those results will be repeatable. Yet what makes science special is that results are repeatable – that is the foundation of science – and hence society can trust them and move forward based on them. How are you, a young scientist at the beginning of your career, going to be able to ensure that your results will stand the test of time and prove to be repeatable?

The best advice we can offer you is 'plan your research thoroughly, carry it out diligently, analyse the results thoroughly, and let the results that emerge guide your conclusions'. A well-designed experiment, done carefully, is much less likely to produce non-repeatable results than a poorly-designed experiment, although even the most carefully designed experiments can occasionally produce results that subsequently prove not to be repeatable. Ideally all experiments should be repeated before any results are published, but often this is not practical. However, it is often possible to design your next experiment based on the results of the previous one in such a way that you assess the repeatability of the results of the first experiment in the second experiment. That way you proceed with confidence; each new experiment builds on the previous experiments. And remember to remain objective when you write up the results of those experiments in a paper (see the section in Chapter 2 entitled 'Objectivity'); be guided by what the results are telling you, rather than by what you and/or your supervisor hoped they would demonstrate. By maintaining a high level of integrity, you can maximise the chances of your results being repeatable.

General points

We realise that many scientists may never do a laboratory experiment, yet still do important research. But that does not mean that this chapter has no relevance for them. Whether or not you are an experimental scientist, adequate planning and good design of a study are crucial to obtain a successful outcome. All scientists need to understand their 'tools', meaning the techniques they use in their research. For example, statisticians need to understand the limitations of their statistical tests, and scientists utilising databases need to know the limitations of those databases, just as much as experimental scientists need to understand the limitations of the techniques they are using. And all scientists need to think about whether or not factors other than those they are studying might be affecting the outcome of their experiments. Finally, make sure that you report your findings honestly and openly – an issue we discuss in Chapter 5. A really well-designed, well-conducted experiment will be of no benefit to science unless the results are reported, and reported fully and honestly.

Checklist for Chapter 7: The basics of doing an experiment

---

1   Start with a hypothesis.

2   Try a preliminary experiment to test the water.

3   Keep it simple.

4   Make your sample size as high as possible and consult a statistician early on.

5   Use appropriate and sufficient controls.

6   Beware of bias and the 'eye of faith'.

References

DeStefano, F. 2007. Vaccines and autism: Evidence does not support a causal association. *Clinical Pharmacology & Therapeutics* 82, 756–759.

Owen, S.F., Huggett, D.B., Hutchinson, T.H., Hetheridge, M.J., McCormack, P., Kinter, L.B., Ericson, J.F., Constantine, L.A. and Sumpter, J.P. 2010. The value of repeating studies and multiple controls: replicated 28-day growth studies of rainbow trout exposed to clofibric acid. *Environmental Toxicology and Chemistry* 29, 2831–2839.

CHAPTER 8

Time management

'Whether we use it or not, it goes' said Philip Larkin in his poem 'Dockery and Son'.

Identify what is critical

At any one time there are likely to be requests from managers, appeals for assistance from colleagues, enquiries about reviewing someone's paper, a need to write a grant proposal, our own inner voice telling us to get on with that paper or to respond to invitations to attend some conference, and that is before we get to the non-scientific distractions. Unfortunately, the older you get, the more numerous these competing demands will become. Each of these calls on your time will have some merit, and doing each and every one could be beneficial to you in one way or another. But, sadly, it will rarely be possible to do them all and you will have to make hard choices. It is naive to think you can accomplish several tasks simultaneously and to the same standard. It is all too easy to be overwhelmed, work long hours, do none of the tasks satisfactorily and feel depressed. You have worked so hard and yet nobody is satisfied, least of all you.

Instead you must identify what is critical according to its value to you (or your organisation) and the proximity to any contracted deadline. This is about assessing what is of the most fundamental importance, sometimes called 'identifying the bottom line'. It sounds simple, but it is not easy. To progress as a scientist, to become more employable, or to successfully remain in employment, what do you need to do? The basics are converting your research into papers and attracting research funding; everything else is a bonus! Whilst this focus should not obliterate all other activities, these aims should remain prominent in your plan for the year. Be aware of deadlines for reports, grant submission dates and abstracts for conferences. Whilst the amount of time you have left before the deadline gives you some room for manoeuvre, we would definitely not recommend starting tasks at the very last minute. Unfortunately, events can emerge which might mean your priorities change and so you will need to adapt and change, too. All this will be considered in your plan.

Make a plan!

Making a plan is vital. This starts with a list of all the tasks which it is either vital or desirable to accomplish. Don't trust to memory that you will keep track of them all. Don't write each task that needs doing on separate pieces of paper scattered around your desk. Next comes the important part. When you have made your list, you then identify the most important of these tasks (prioritising). In my case I either put numbers (1, 2, and 3) or stars beside them. This is the critical stage where you have analysed what is vital and/or close to a deadline. These are the tasks you will focus on as essential. My recommendation is to keep a notepad by your computer where all the tasks are listed. Alternatively, these could be written on whiteboards or on a sheet of paper pinned to your wall in front of you. We realise that personal electronic devices or your PC offer advantages in keeping lists, but they are dependent on being switched on and your remembering to look at them; those lists can very easily disappear off the bottom of the screen in front of you!

Now relax!

You have made your plan, you gave it some thought and have decided on your priorities. You can do no more; there is only a finite amount of time in the day. Not everything on that list may be completed but you have identified what is vital. Don't forget the most wonderful and satisfying part of any list is putting a line through things when you have completed them! The written plan has other advantages. If a manager rushes into your office with yet another task for you to perform, you can alert them to your list and enquire which task you should now drop in order to help them? Similarly, your line manager will be reassured to see your prioritised list in a prominent place in your office. Even if you are not there, he or she can see your thinking and be comforted that things are not forgotten.

I find wall planners are also very helpful. Here you can flag up those deadlines and also try to carve out and block time (with marker pens or stickers) which you will dedicate to your paper or experiment in the lab. It helps you and others to understand your schedule should they come to your desk/office. Sometimes, just before I leave my desk at the end of the day, I will put a little note on the keyboard to say 'Start with X', or 'don't forget your meeting at 10 am'! This ensures you start the new day on the right track.

Clarify what needs to be done (double check)

Every type of human interaction at every level will always be plagued by ambiguity and misunderstanding. Whilst person A feels they have made their request perfectly plain to person B (and therefore they don't need to repeat it or express their requirement in another way), they may be amazed to discover that person B has interpreted their request in a completely different way to that intended. Whilst person B may well have grasped that a piece of work is urgent, they may still end up working furiously on the right topic but from the wrong direction. This situation is far more common than we admit and is a major cause of inefficiency in any organisation. Therefore, it is very wise for all parties to ensure, and double check, that they understand one another. That is, both parties should repeat and clarify what it is that needs to be done, how and by when. Once the task is started and some progress made, there is again an opportunity to compare notes and ensure things are going in the right direction (before it is too late to change). Never trust that 'it is plain obvious'!

Protecting your time

Beware of time thieves (I have heard them called 'time vampires') that will unwittingly steal your valuable time. A convivial lunch or extra-long tea break may help to raise your morale but disrupt your schedule. Distractions can include the friend needing advice on their love-life or advice on getting their complex piece of equipment to work, both of which can be extremely complicated! The latest gossip on office moves, resignations or strategy documents from headquarters will all have their fascination but they are nibbling away at your time. Whilst we would not advise you to shut yourself away from society, there will be times when you must be ruthless and explain to your colleague that you have to 'get on'. Sadly, your chatty friends won't be there beside you when it is necessary to explain to a manager why you have not completed a task. Failing to complete your paper for another year affects your career and not theirs.

We will be bombarded by email from the moment we switch on our computer at the beginning of the day till last thing at night. Many of these emails have merit and seem to require a rapid response, no matter how trivial the issue. What's worse, you know that tomorrow there will be many more emails. But if you compare those email requests to that list of tasks written down beside you, ask yourself 'are they really of equal merit?' I sometimes think trying to focus on a task with emails pinging in is like trying to paint a beautiful picture with wasps buzzing round your head! So consider limiting looking at emails to an hour either in the morning or after lunch, but other than in those periods, keep your email resolutely closed! The same view might be taken of social media, something we discuss in Chapter 13. The attractions of social media can be terribly beguiling. You can learn many new things and have enjoyable interactions with colleagues all around the world. But time is leaking away at a rate that sometimes seems impossible to believe. So again, our advice would be to set yourself a time limit.

Given these various demands, another way to think about defending your time and maintaining your productivity is to ask yourself 'what time of the day am I most effective and performing at my best?' If the answer is 'in the morning', then leave the slightly less important or less demanding tasks till after lunch. I once asked a senior colleague of mine how he managed to both run a department and write five or more papers each year? He answered that he preserved a 'golden hour' each day. Every hour before 9 am was truly his and his alone. Then he would work on his favoured task (usually a paper) for an hour or more before the phone started ringing!

Preserving your sanity and your mind!

It is tempting, as that list of tasks grows ever larger, to work longer and longer hours in the hope that you will inevitably see the 'light at the end of the tunnel'. In other words, you think taking on more tasks will not be fatal, as you can simply work longer hours. After all, there is nothing to stop you working through much of the night and the weekend too? But there are several things wrong with this strategy:

- The quality of the work you produce will suffer as your mind becomes tired.
- Your personal and family life will suffer.
- You lose thinking time.
- It is unlikely things will actually get better anyway.

I have been struck by how many successful business leaders put a lot of their success to simply saying no! That is, they prioritise their efforts in the few key areas that really matter and at the same time have the courage to say no, and disappoint people. They have learnt to focus only on the important issues.

Devoting time to people, meetings and conferences?

Whilst we would encourage you to maintain a streak of ruthlessness and focus on the critical tasks, there remains a balance to be struck with maintaining your professional relationships. Indeed, you should always remain on the lookout for interesting people with new ideas. Taking time to seek them out and have a conversation could stimulate your mind into seeking new approaches to scientific problems (so you achieve your goals quicker). If a speaker has given a talk, either in your department or at a conference, that impressed you, then try to approach them afterwards and draw them out in conversation.

We don't believe successful project management can be achieved purely by firing off brief emails. Much of science succeeds through partnerships as people of different skills combine. These partnerships need to be nurtured; in other words, do invest your precious time in them! If you have to attend a meeting, do ensure there is an agenda and ensure that minutes are taken. If you are organising the meeting yourself, then work hard on producing a clear agenda beforehand and keep the meeting short and to the point.

If you are in demand for several informal or formal meetings, don't organise them in your diary as starting back to back. This may seem an efficient way of getting them out of the way but you end up staggering, poorly prepared, from one meeting to another.

If an ongoing project or new consortium enquires about you helping them by providing a scientific service, think carefully about the time and money this would actually take. Don't be over optimistic. Many would argue you should 'take your worst estimate and double it!' It is better if your colleagues are surprised by you producing the goods ahead of time and under budget than the other way round.

Your character and time management

Whilst we like to think many of the techniques of good time management can be taught, many issues and troubles might be linked to a person's character. Sending staff off on time management courses may appear to benefit an individual for a couple of weeks but things usually return to their previous unhappy state of affairs. A former colleague of mine frequently explained to me that they simply 'ran out of time'. The problem was with me (giving them too much work), or with the world in general. The failure to complete tasks was always due to external forces. They entertained a hope that in future things would become a lot quieter and then all tasks would finally be completed. Sadly, I knew this Promised Land would never arrive! So perhaps a time comes when you need to be honest with yourself! For example, are you an:

- Incurable optimist?
- A never satisfied perfectionist who never completes?
- Someone who simply does not gauge time?
- Someone incapable of prioritising?

Try to understand your character. Could your personality be part of the problem? Ask yourself if there are any approaches that might help you personally. If you know the required end-date, then work backwards and ask yourself how much effort can you devote before time is up? Don't let the perfect be the enemy of the good.

Give yourself a break!

We hope that reading this chapter has not raised your stress levels beyond what they were to start with! Whilst a level of cold efficiency is necessary, you are not a machine. If grinding efficiency was all that was necessary, scientists would be on the list of professions due to be replaced by robots too!

We should treasure scientists for their curiosity and independent spirit. You need opportunities to be imaginative and try new things. So it is valuable to step off the treadmill now and again and look around. Keep some curiosity time for yourself. Maybe read a new research paper that is not in your field but seems interesting. Look out of the window, take a walk and allow your mind to think freely. When you have gathered your thoughts, have your colleagues round for tea and biscuits, and try to explain your ideas to them and get them to challenge you. Don't succumb to the view that ideas come from 'brain storms'. Good new ideas emerge from months or years of mulling over a problem. Now where's that kettle?

Checklist for Chapter 8: Time management

1   Identify which tasks are critical and prioritise them.

2   Do make a plan and keep it near you, in plain sight.

3   When working with others, double check that the task is understood by all parties.

4   Protect yourself from the time vampires!

5   Keep nurturing scientific relationships.

6   Check if your character is playing a detrimental role.

7   Permit yourself breaks.

CHAPTER 9

Giving a presentation or a poster

The need to communicate well in science is not appreciated as much as it should be. Some feel that their duty as a professional scientist is only to churn out good results, and it is up to the rest of the world to recognise this and try to keep up with them. Unfortunately, good science can then go unnoticed by the wider scientific community, due to the failure of scientists to explain what they are doing in an illuminating and memorable way. Giving a presentation or poster is often the first test of ability to communicate that the young scientist will face. But these are also skills that we suggest all scientists should continue to review and strive to improve. Fortunately, quite a few departments and businesses offer training courses, particularly on giving good presentations, and you should definitely take up these opportunities. Here we will offer a broad view of the background to giving good talks and posters, plus some technical suggestions.

Step 1: Understand that a presentation or poster is a very different beast from a report!

What we mean by a report is a formal written document explaining in minute detail what methods were used, and including all the different results and analyses carried out in a project over a number of years. This is something organisations typically have to do to explain to the sponsor of a project what they have been getting for their money. A PhD thesis is a similarly thorough report of all the work done over three or more years. With a report or thesis, the reader is at leisure to read the information (or not) at their own pace. Unfortunately, many people assume that it is their duty to cram as much of their report methods, data and graphs into their talk or poster as they possibly can. Bizarrely, we have even witnessed people's second slide functioning like a contents page in a book, saying their talk will have an introduction, methods, results, discussion and conclusions sections! So why do we insist that a talk or poster is different from a report? It is because the audience only has minutes to follow the work you did and understand the significance of your results. Not only that, but your audience is likely to have minds already burdened with trying to recall the many preceding talks and posters they have listened to or witnessed during the day. So you must acknowledge the limitations of the human capacity to absorb information in the few minutes available to them when confronted with your talk or poster. Knowing this to be the case, we must approach this task in a different way to a written report.

Step 2: The critical steps required in your thinking to communicate well

Your first ambition should be to communicate with every single soul in your audience:

> *Your job is to ensure everyone can follow your talk, no matter what their background.*

Unfortunately, there is a tendency for presenters to prepare a talk in a way that placates the probably small section of an audience who are already very familiar with their topic. Perhaps they see in their minds' eyes some particularly grumpy group of professors, or unsympathetic rivals working in the same field. They feel over-anxious about potential criticism and so their instinct is to obliterate it with a barrage of complexities and technicalities. However, this attempt to deflect potential criticism won't succeed on any level. First, because you will never change the minds of grumpy people and second, because you have not reached out for the support of the wider audience; in fact, you have probably now alienated them.

To turn this around, the most important philosophical step you will need to take if you are to give a good presentation or a poster is to think deeply about the needs of your audience first. This is the big mistake that many people make when they believe their duty in giving a talk or a poster is simply to tell people what they know, and perhaps all they know, on a topic. This is like a sort of dump of data downloaded from the internet. The audience must try to keep up as best they can. If they can't keep up, that's just too bad!

A really good speaker will actually do quite a bit of research into just who their audience will be, their level of understanding of a topic and their concerns. Thus, the more inexperienced an audience, the more effort must go into the introduction. In other words, your topic might need a longer explanatory lead-in than you are used to. Also, in these circumstances you should review your vocabulary carefully to replace terms which are common knowledge only to your circle of experts. It is increasingly common for you to be in countries with different cultures and comprehension of English language to your own. So don't be so ambitious in trying to show too many slides with lots of words to this audience. Let the pictures and graphs do more of the work!

Your aim should be:

Tell people only what they need and want to know.

This means considering what information the woman or man in the street needs or wants from your project. Thinking this through requires a bigger intellectual step on your part than at first seems necessary. Essentially, what this requires you to do is to imagine you have now become someone else who knows nothing about your topic, but is curious. Admittedly, this is very hard to do! The problem is that you have been thinking, working or even dreaming about your research topic intensely for many months, if not years. Your research project has become second nature to you. You understand completely the importance of the topic, why it was first considered, the main weaknesses in previous studies, etc. So you see no obvious need to explain this to your audience. But it is clearly wrong to assume your audience has all studied the same subject as intensely as you have. That is why you must try to take a step back and view your subject as an outsider who has no prior knowledge might. Ask yourself what sort of questions might be going through the mind of this outsider.

What would an outsider wish to know on first hearing of your topic? The questions this outsider would ask will revolve around the words 'why' and 'so what'.

- Why that topic?
- Why those chemicals, species, rocks, etc.?
- Why there?
- Why now?
- What is the innovative step you have taken?
- What can you conclude?
- Why might the result matter to me now or in the future?

The second step you have to take is to:

Be sympathetic to the weaknesses and limitations of your audience

Despite many of us spending a considerable time during our careers sitting in an audience listening to science talks, few of us allow for the reality of people's limited capacity to actually absorb new information. What do we mean by the limited capacity of the audience? Surely they will mostly be super-knowledgeable professors, brim full of tricky questions? What are these limitations you speak of? If you have attended any conferences, then it is likely you have experienced them; they include:

Tiredness – Most people will start to suffer from their minds wandering (your talk may be one of several that day), and they will easily lose the thread of what you are trying to say. After a few of your more complex methodological slides, often peppered with a few equations, you are likely to lose them entirely for the rest of the talk.

English comprehension – Neither you nor your audience may be confident or competent in English. This is certainly not the fault of the audience! You should have as a principle a desire to reduce the numbers of words and text to a minimum, so that the audience can focus on the bare essentials. Also, speaking at the rate of a breathless sports commentator (as you try to cram it all in) won't help native, let alone non-native, speakers follow your study. We advise using pictures to illustrate most of what you are saying. It's an old saying but it's true, 'a picture can be worth a thousand words'.

Depth of knowledge of the audience – Don't assume the entire audience is already highly knowledgeable about your topic, understands the background completely and is familiar with all the specialist acronyms used in your field. All too often, people dive into the complexities of their research and do not allow for this lack of background knowledge of their audience. OK, there might be specialists present, but perhaps those who are curious or intrigued by the field have wandered in to hear your talk. There really is no harm in taking your time to give a gentle and sympathetic introduction, so that everyone is brought up to speed.

Speed of thought – We are likely to have become quite immersed in a topic as we study it relentlessly for month after month, year after year. Yet we give our audience between 20 and 60 seconds to grasp what has become second nature to us! Be kind to your audience; think of how little you knew at the beginning! So try to keep your slides clean, with simple messages. Don't clutter them up with several graphs, figures, many-coloured texts and ambiguous wording.

Eyesight – You might be surprised at just how little people can see from the back of a conference room. Our assumptions about the ability of someone to rapidly grasp the salient points from our many graphs in a few seconds from the back of a room are typically over-estimated. Many presenters take their tiny, detailed, complex graphs and tables straight from their report (yes, that sorry report rears its head again) or paper straight into their talk. How many times have we heard a speaker say 'I'm sorry if this is not very clear', or 'you probably can't read this from the back – but!' Well, in that case you should not have used it! Take time to increase the font size and clarify, or don't use it.

Memory loss – The most common situation is that you will be giving one talk amongst many during the day. Perhaps an audience could be sitting through ten talks over the course of a day. Now imagine each talk ends with five bullet point conclusions. How likely is it that at the end of the day you will be able to recall all 50 of them? Yet, naturally, we would like people to remember something of our talk. Clearly you will have to scale back your ambitions on all those many conclusions you might like to raise. So, if there was only one thing you wanted your audience to remember, which one would it be?

Considering the many limitations faced by this all-too-human audience in following and remembering anything of your talk seems at first rather disheartening. But if you acknowledge these weaknesses, there are aspects of human nature you can actually turn to your advantage. These include our love of stories, our quick response to visual images, together with engaging with the audience as fellow human beings.

Step 3: Utilise the human desire to be told a story

Try to think of turning your science results into a story, like one you might tell to recently acquired companions around a campfire in the woods. Your story should have a form like this:

- Explain why the topic is important and then how our absence of knowledge/suitable techniques is holding up science.

- Then you, the new scientist, walk into town with a novel bag of tricks to tackle the topic. Explain something, but not too much, about the techniques and experimental design. Hopefully, you have now generated some curiosity and suspense about what the outcome was.

- Select some (but not all) of your most crystal clear results to show just enough to complete that one clear story.
- Deliver a user-friendly, satisfactory ending.

As discussed above, your natural instinct in a talk is to tell an audience everything you know, roughly following the chronological order of your project.

Avoid doing so!

As a story teller, you select the minimum number of slides that will tell a satisfying story. This talk may cover only one part of your bigger project. Yet that story should reflect and give people an indication of the wider work you have been doing. Don't worry that you don't give them absolutely everything. If your story is a satisfying one, then at least some in the audience will remember you, the author, and look forward to your next story! Those that want more detail can come and ask you afterwards. Remember, keep the number of slides to a minimum and cover one, not five, points on each slide. Don't worry, you won't be selling yourself short.

Step 4: Having good slides and a dash of honesty

We will discuss ways of presenting scientific data in the next chapter, on writing a paper. Thus, hopefully you can lift figures from your papers, talks and posters, and vice versa. Remember, we talked about your audience probably being a little tired, as there could be many talks over the day or several days of a conference. So how can we help these tired brains understand your slides?

- Our first tip is to use pictures wherever you can! Perhaps due to the origin of our species, way back in time as hunters on the African savannah, we humans pick up and process images very quickly indeed. So try to scale back on words, and instead illustrate with pictures. You could use pictures to help explain concepts or simply double up on your words (the illustrative picture can sit alongside the word). This then helps you in two ways – it speeds up comprehension and the colour brings life to your talk, almost like a blood transfusion! Imagine a series of slides filled only with words and, if you are lucky, a few tables. Pretty dull. Doubly and fatally dull if you are tired, have had a good lunch and are sitting in a warm room!

- Our second tip is to use text boxes and arrows to highlight the key features of your figures. Whilst some might call this 'spoon feeding', it is necessary because of the short time available for people to absorb new information. For example, a speaker may show a graph, which is an important part of their story, assuming the significance will be blindingly obvious to the audience in the few seconds available for them to examine it. But you shouldn't rely on this! So we recommend adding a text box either beside or in your figure, perhaps as a balloon with an arrow pointing to that critical point in your graph, where you state simply what has happened that is significant.

- Our third tip is to show your human side. As humans we can all connect and empathise with the doubts, fears, tears and joys of research. Add one or two titles or summary statements to your slides in everyday speech. These might be *'My disastrous first experiments'*, or *'These results were totally unexpected'*, or *'Surely these results could not be right'*. Here you are speaking as yourself, revealing your own first reactions. This is not for every slide, but you are reaching out to the audience with everyday language; you are one of them; you are cutting through all the mysterious jargon and showing your human side. Forming these human bonds will help draw your audience to you.

Step 5: Review your confidence and motivation

It is natural to feel some anxiety or nervousness before giving a talk. You are up on a stage and in front of many pairs of eyes. As a beginner, it can be a real struggle to overcome these feelings of nervousness. You might wake in the middle of the night from a nightmare of being on a stage, forgetting your talk and muddling up your slides, as your exasperated head of department shakes his or her head in despair! Don't think like this. The audience will typically want the speaker to do well.

Is it an exam?

No, you should see it as an opportunity you should grasp. In fact, it is a privilege. You now possess a lot of knowledge that others don't have. You have a project; it is your baby and you want others to admire it! It is a privilege because in this day and age most people are stressed and very short of time, so getting them to stop and listen is a rare occurrence. They are surrounded by devices clamouring for attention or demands from their superiors to produce that report they promised. So if people take time out and set aside their daily demands, that is an honour you must respect. Your aim should be to reward and inform your audience rather than punish them and make them miserable!

We find the best way to boost confidence before giving a talk is simply by being in the audience and listening to the preceding speakers (assuming you are not the first speaker at a meeting). You then get a feel for the atmosphere and a feel for the audience. In a small way, you are just starting to feel a little at home and know what to expect. But often you can feel that other speakers are not so brilliant; in fact you could do better!

Step 6: Practise beforehand

Preparing your talk is a silent and lonely business. You convert your ideas and graphs into slides on the computer before you. You have a rough idea of what you want to say and often you produce an overabundance of slides for the amount of time you have available for your talk. Converting this into a living and breathing presentation means you have to practise giving the talk and getting the timing right. I often start the process by switching to 'slide show', putting on a stopwatch and speaking out loud to the computer. Then it is a good idea to practise in front of friends and, potentially, your supervisor. This will often show that you have too many slides and are trying to cram in too much information. A good rule of thumb is one minute per slide. So a 15 minute talk means only 15 slides. Go slowly when it comes to talking about your graphs, and start by explaining what the axes are and their units. Memorise the key issues that you want to bring out in each slide. The main benefit of practising is that you will be a lot more confident of doing yourself justice when you actually do get up to speak

Step 7: Your style and behaviour on stage can make a difference

Familiarise yourself with the room, check and test your talk in advance

Before your talk, try to go up to the stage or front of the room where you will speak. Get familiar with the set up and, if necessary, rearrange it to suit yourself. The podium or laptop should be to one side of the screen. Ideally the podium/ laptop should be at 45 degrees to the audience. Check and adjust the room lighting if you can. A totally dark room is good for slides but encourages sleeping in the audience and increases the likelihood of you tripping over a cable! A bright room can make the slides look indistinct, so try to find the best compromise. You should be able to look at the laptop screen, look across easily to the big screen itself, and look at the audience without difficulty. There are now two things to consider. How will you use your voice and your personal movements on stage? If you are provided with a microphone, perhaps with someone's help sitting at the back, check how it sounds. How far should your mouth be from the microphone? Do you need to raise your voice or keep it at an ordinary volume? If there is a laser pointer provided, check you know how it works (without blinding yourself).

Use your voice to add some drama!

When giving a talk, try not to speak in a monotone! Normally, at least in our scientific writing, we try to keep to an unemotional style. But remember you are now a story teller and your job is to keep people awake and interested in your study! After all, when you look at your slides and results they do not all have equal importance. You should raise or strengthen your voice and gaze directly at the audience at the most important moments. These might be the big unanswered questions you reveal at the start of the talk. Perhaps it is the amazing opportunity of trying out this new analytical instrument or your surprise or dismay at the first results. Then end with some provocative conclusions. You felt emotions at these moments in your research and you will help yourself as a communicator if you show some of this emotion to the audience too. Some scientists (those who are more conservative) may see any such attempts to dramatise your story as surely a case of being a phony actor! Not true! You care about your science, you are keen to communicate, you want to hear others' opinions and you won't help yourself if you speak in a quiet, unemotional monotone.

Move a little

If you can, it is good to make some movements on the stage. A natural move is to go to the big screen and point to the axes of a graph and explain the units. Perhaps also move to the screen to highlight the most exciting part of your results, pointing to a particular curve on the graph. This movement will help in two ways. First, it adds a bit of life and drama to your talk, it draws the eye and shows the audience you really care, you really want to ensure they understand. Second, it can help to relax any tenseness in your muscles. Standing rock still at a podium is not how you would normally communicate to your friends. It makes you feel and appear cautious and anxious to the audience.

Focus on the audience

Try to avoid, at all costs, carrying a script with you on stage, or notes to your PowerPoint talk. You should be able to remember what you want to say. But you don't need to memorise your speech word for word; the same story may be told in many different ways. Your gaze can now be shared between the screen and the audience (rather than being distracted by written notes). As far as possible, keep your body turned towards the audience. Whilst I recommend walking towards and looking at the image on the big screen, do this side on, never turn your back on the audience. If your back is to the audience, then who are you giving the talk to? Look at your audience but don't focus on only one group or individual! Remember, a room has four corners and you must not let anyone feel left out or not a valued part of your audience. If you watch TV, you will note that most politicians are very good at this! You are now a communicator, it matters to you that people understand, people should see in your face that it matters, that you want people to engage. They should see this in your eyes. If your face maintains an utterly flat and unemotional demeanour it disengages people. If you don't seem excited by your results and work, perhaps they shouldn't be either? Remember we are all humans. As you talk to your friends at home or at work and explain what excites you or what is going disastrously wrong, they can see these feelings in your face. So treat the audience as a new group of friends you want to reach out to.

Things to watch out for

A list of common mistakes

- Not keeping to time.
- Using text that not everyone will understand.
- Not clearly labelling the axes and their units in graphs.
- Tables that cannot be read beyond the front row.

- Plunging into lots of unnecessary methodological detail.
- Assuming the audience are all experts and practitioners in your field.
- Not looking at the audience.
- Talking in a monotone.
- Overuse of animations.
- Non-existent or unclear conclusions.

Poor talks with the wrong sort of motivation, given by more senior speakers who should have known better include:

- Over-selling their expertise

  Here we have a naked marketing pitch. They blitz the audience with impressive models or instrumentation they possess. They don't worry overmuch about conclusions, just showing the extraordinary potential and expertise they possess and you so sadly lack. Other scientists can only feel despair and insignificance whilst potential funders leap to their feet in excitement!

- Battering their competitors

  Here the speaker uses hundreds of slides chock full of esoteric acronyms with little in the way of conclusions. They are saying 'I have done or am doing everything so there is no need for you to even bother turning up to your lab tomorrow!'

- Arrogance

  I have been doing this stuff for years. Yes it is complex, but surely you have been keeping up with my many papers in the past? Where have you been? If you have to have it explained to you, then you are clearly a somewhat inferior, lower level of scientist! What are you doing at this meeting anyway?

The ideals of a talk in brief

- Discover the length of time available for your talk and use the minimum number of slides. A comfortable pace is normally one slide per minute.

- It contains a good story.

- A well-practised talk will come over as professional and show people you care about them and about getting it right.

- Avoid, if not eliminate entirely, the use of acronyms.

- Have one clear 'take home' message.

A last thought - hopefully, you will get lots of opportunities to give talks and maybe even start to enjoy them! Now this issue is a little sensitive, so don't take it the wrong way – do you have any irritating mannerisms? Our best friends will know that we have these but they may be too polite to mention them to us. So perhaps to get that final polish you could ask them for their honest observations? We may find it impossible to eliminate them entirely. But if our aim is to continually improve, then no stone should be left unturned.

Checklist for Chapter 9 – Part A: Giving a presentation

1   Be sympathetic to your audience! Don't overestimate the knowledge of your audience or how quickly they can assimilate your information.

2   Consider your talk as a story and draw your audience to an inescapable conclusion.

3   Manage the amount of information presented on each slide wisely so that it is easily digestible.

4   Ensure all material is legible, even from the back of a large hall.

5   Always strive to help your audience understand by carefully identifying the key findings.

6   Avoid acronyms as much as possible.

7   Show honesty and your human side.

8   Look out at your audience so that no one feels left out, some movement will help you feel relaxed and keep the audience with you.

9   Practise and ensure you can keep to time.

Presenting a poster

Posters can be a nuisance to prepare and somewhat dispiriting to stand beside as people ignore you to focus on their canapes and on old friends from university. Alternatively, they can lead to some of your most meaningful discussions, develop great contacts and be surprisingly influential if things go well. To some, a poster has served its purpose simply as a marker to say, 'Hey guys, we are working in this field so watch out'. Some use a poster as a form of comprehensive report with vast quantities of text and figures on offer. But unlike papers and presentations, you have a remarkable degree of freedom to design your poster as you see fit. So you are in the unusual position of being able to enjoy some artistic creativity! However, note that a poster session is, perhaps more clearly than other types of science outputs, a form of competition. In a subtle way you are one of the traders in a covered market trying to attract the punters. Can you attract and hold their attention sufficiently to sell them your science goods?

As you are reading this book we assume you are striving to become a better scientist and achieve excellence in all things, so let us investigate what might be the ingredients of a really successful poster? Just as with a talk, the best way to start is to consider the limitations of your audience and their constraints.

Acknowledging the weaknesses and limitations of the poster viewers

Lack of time

Typically, poster sessions are organised to fit in around the breaks between the oral sessions in a conference, such as at lunch, tea-breaks or at the end of the day. At large conferences there may be a few hundred posters on display simultaneously and then only for a short time. So your audience will be trying to get something to eat and drink, as well as find an old friend or person they need to lobby, before the next oral session begins. This, unfortunately, shortens the actual time available for poster viewing. With a poster session at the end of a long day many feel they are exhausted and have little energy or time left to devote to posters.

A tired audience

Having sat through several talks, they could well be rather jaded and their minds distracted by the many, hopefully interesting, oral sessions they took part in. So, just as we discussed with giving a talk, you have to accept the reality that your audience will be somewhat tired and finding it difficult to focus.

Difficult to see through the crowd

Another not uncommon problem is crowding. When two or three people stand in front of your poster to read it, they shield much of the text from the casual passer-by.

Those on a mission

It is quite usual for people at a conference to have already made a note of the posters they must see, due to their potential relevance to their own work. They will make a bee-line for these posters and dedicate most of their time to them. But as they wind their way around the posters whilst trying to find the ones they have targeted, they might pass your effort. Is there something about your work that might catch the eye and be easy to digest?

So what should we conclude from all this? MAKE IT EASY FOR THEM! There are some scientists whose attitude appears to be that their poster is designed to communicate with only the privileged few technical specialists in their area at the conference. But this is a missed opportunity. Good scientists will always want to disseminate their work and findings to as many people as possible. This probably won't alienate the technical specialists but the other approach will surely minimise the amount of impact you can make. The ideal is that your poster will provoke people to ask you some serious and interesting science questions, rather than you having to take a lot of time explaining your poster (due to its small text and complicated figures).

Responding with your poster design

Title

Keep it as short as possible! You want the passer-by to quickly take in what you are about and perhaps feel that what is on offer will be a tasty 'scientific snack'. In addition, long titles will take up valuable space on your poster.

We would recommend including an image of yourself (and perhaps a co-author or two with you). Although this may seem a bit personal (something we generally don't go in for in science) or cheesy, this can really help spectators. That way they can easily spot in a crowd who is the author and so ask them their questions about the poster. It can be very frustrating to have a burning question and find either that the author is not obviously near his or her poster or that she or he is possibly buried somewhere in a scrum of people. All you have is a name, or several names, to hunt for on name badges during the rest of the conference. So it is more than likely you will give up. Don't forget to add your email to your poster so people can contact you with questions or a request for a copy of the poster.

Background colour

You have a lot of choice and we don't want you to feel inhibited, but we have found that white is a reliable background colour. Your text and figures will show clearly from a distance. A very pale background colour can also be successful, such as when you separate your poster into a series of boxes which can contain text and figures. However, darker background colours, such as green or brown with a black text seem less distinct and somehow have a rather depressing effect. You could put white text on darker backgrounds, such as blue, but we don't think these are so successful.

Text size

Make sure the text is clearly legible from at least one metre away. Your text will be an important part of the poster and if it is a real struggle to read, then it is likely your potential viewers will quickly move to another one that is easier to follow! No one will complain if your words are too big but you will get complaints if they are too small. It is customary to have handouts of your poster available (don't forget to bring your business cards too). Choose a font size that can be printed and read even when your poster is reduced to a handout size (A4 or US letter size paper sheet). Depending on your software and printer, it may work to design the poster for your handout size and simply print it larger when finished (but check that pictures and graphs have sufficient resolution).

Organisation

If you have a choice, we would recommend portrait orientation. It is generally easier for the eye to move from top to bottom than take the up, down and sideways route often necessary in a landscape poster. You will notice this problem particularly where the work in a landscape poster has been organised into three, four or more columns. Some might argue that a landscape poster, perhaps higher up on the board, is less likely to be blocked from view by spectators. However, this will be nullified if the organisation of the poster makes following the thread of the story hard work.

Make an easy route to follow for the eyes. We would normally read from left to right and then work our way from top to bottom. With a series of columns, we must break off and re-find the start of the next column. This is not straightforward if there is a jumble of figures and subheadings competing for your attention. With more than two columns, the route that the eye has to take can become something of a confusing labyrinth. This could lead to fatigue, confusion and the spectator feeling defeated. He or she then simply moves on to the next poster.

If you have your work structured in two or more columns, ensure that each new column starts at the top with a new heading, rather than being a continuation of the text at the bottom of the last column. Otherwise such breaks will make it hard for the eye and brain to keep following the narrative.

There are a number of ways you can help guide the eye through the story of your poster:

- Make it one column, so it is clear the eye has only to move from top to bottom.
- Put the text and stages of the poster, e.g. introduction, methods etc. into a series of discrete but clearly outlined boxes, perhaps with arrows guiding you from box to box.
- Number the different stages of your poster story in sequence.
- Use standard, recognised titles such as introduction, methods, results, conclusions.

Content

Try to turn it into a story

If there is a way you could turn your work into a satisfying story it will massively increase the chance that people will remember your message, and you too! As we mentioned before, humans like stories! A good story will leave them with a warm feeling and a measure of gratitude to you. A successful poster will not need you to be there to convey the essential message but will still be so intriguing that the reader will want to ask you questions.

Don't use too much text

Posters are not the right place for lots of text. It takes a lot of time to read and interpret large blocks of text. Lots of text usually means small text, which is also hard to read. The casual passer-by will look at your poster and think – 'that looks like a lot of hard work to read and understand, and I don't have time'.

Some brave souls make do with very little text indeed. This could be extremely successful in getting the key essentials over to an audience in a short space of time. But it would need to be extremely well thought-through. Is the minimal explanation still suitable for a wide, non-specialist audience? So there is the opposite danger of leaving the reader mystified because you did not help explain enough of what you did, why you did it and the value of the findings.

Balancing your text, images and figures

It is worthwhile spending a lot of time finding the right balance between text, images and figures. We would recommend not using more than two (even one is preferable) of your data graphs. Pick the key one(s) and these will be the foundation of your story. You can use tables too, but they are terribly dull for the eye, so better to avoid them or use small ones. Images are good for bringing your topic and points to life but not too many otherwise you risk cluttering up the poster and confusing the route the eye has to follow. The text you use should be the minimum required and absolutely crystal clear and unambiguous. You really have to nail it! Avoid acronyms and long sentences.

Maximise your impact

Normally, we would advise you not to exaggerate or hype up your work. However, you still need to find ways to make your work grab the attention of the casual passer-by. Try to get an early win with a provocative/intriguing introduction, possibly with a simple image or figure high up on your poster (so somewhat above the many of the heads of the crowd) which is very straightforward to understand. Hopefully, you will get the balance right between good science and a simple story which has a potentially big impact. Yes, keep spelling out the impact – what does your result mean for a wide audience?

Conclusions section

Always aim to have some! We recommend the word 'conclusions' over alternatives such as summary, key observations, major findings etc. because conclusions is a universally understood word in science. Make the conclusions meaningful to a wider audience, not just a small group of technical specialists. For example say 'these results suggest: if current trends continue, the frog will be extinct in two years', rather than 'the work shows the model works in a reasonably reliable way to predict frog trends'.

As people walk down a long parade of posters they may only look at the conclusions section to check whether you have found anything interesting. Then depending on how clear, interesting or valuable your conclusions are, they might dedicate more time to your poster. If you only state 'the model had an acceptable performance but can be improved with more research' they are unlikely to give your poster a second glance.

Finally, we recommend you don't put your poster into the hold of an airplane as baggage on a flight. Instead, keep it as hand luggage if at all possible. A number of things can go wrong with baggage in the hold. Please remember not to leave your poster in the airport toilets as you board your flight! We know of at least one example of that happening.

Now stand by your poster!

The presenter should do everything in their power to stand by their poster during 'poster sessions'. Not doing so might give the impression that the presenter is either too proud to mix with ordinary punters or, alternatively, embarrassed by their own work? The reader might think 'if the presenter won't make the effort to stand by their work why should I bother?' If you have done your work well, many readers might be very keen to talk to you. Often, you can learn a lot from such discussions. Do not waste this golden opportunity!

Checklist for Chapter 9 – Part B: Giving a poster

1   Be sympathetic to your audience! Don't overestimate their knowledge or how quickly they can assimilate the information.

2   Use the minimum of text, images and figures – enough to attract interest but not so much that your poster looks hard to follow. A poster is not a paper or lengthy project report!

3   Consider the route that you want the eye to follow. Make it a smooth progression without too many jumps and breaks.

4   Try to describe your results in the form of a story.

5   Avoid acronyms as much as possible.

6   Ensure all material is legible even from 1 m away.

7   Always strive to help your audience understand by carefully highlighting the key findings.

8   Avoid complexity, construct your work as a story and draw your audience to an inescapable conclusion. Could the poster be understood in 40 seconds?

CHAPTER 10

Writing a (good) scientific paper

Although in this chapter we focus on discussing how to write good papers, there is much here you could consider as good practice in any scientific document you have to write! Actually, you can now find many excellent books and web resources on this topic other than in our chapter. Fortunately, you will find we are all giving a similar message, although some advice might have a slightly different emphasis.

Introducing the concept of a good paper

The heart and soul of a good scientific paper is the quality of the science and thinking that went into it. From reading this book you will have an impression of what we mean by good science. Hopefully, you used hypotheses, you planned your experiments carefully, and you demonstrated objectivity and acted with integrity? Now what is needed is to translate all this good science into a paper! What do we mean by good paper? We mean one that will grab the attention of scientists in your own field but potentially also be noticed by scientists in other disciplines. If you succeed, this will be evident over time, as it feeds through to the citations your paper earns over time. Citations electronically accumulate as others cite your paper. This is the best evidence we have that a scientific impact has been made. However, you will find variations in the rate of citations of a paper across different disciplines according to the size of its field. Thus, a good paper on human health is likely to achieve many more citations than a similarly good paper on, say, geology.

In theory, writing a paper is merely transferring your knowledge using impersonal language and a few figures into a manuscript. As you are communicating with fellow scientists, trained like yourself, you might expect this paper writing business to be fairly simple and mundane. You might think that all scientists swiftly appreciate the scientific merit of all science papers in their field. However, this is not so. Far too many papers fail to have an impact and indeed fail to be understood by their intended audience. So this chapter will consider how best to communicate with the scientific community and so improve your chances of your work getting noticed and having impact.

Starting with the right mind-set

When we think about what distinguishes good authors who regularly write influential papers, they seem to have a number of valuable characteristics, such as being well organised, clear and honest, with an arrow-like focus. But probably the most valuable characteristic is a real passion to communicate. You have to really have a desire to reach out to the scientific community. This passion is needed so that you are continually asking yourself 'am I getting through to people, will they understand?'

The language of science

There is a common misconception that the sign of a clever scientist is one who communicates in their papers through the use of highly complex scientific language. Thus, lesser mortals have to read the sentence several times before understanding it and even then, they probably have to refer to a dictionary (or search the web) along the way. Sometimes we feel that because we have not understood the complexity of what is being said, it must surely be wonderful science, certainly above our modest level of intelligence. But in the authors' view the opposite is more likely to be the case! Good scientists write in a simple and clear way because they have that desire to communicate to as wide a range of people as possible. Some have defined 'plain English' as a sentence that can be readily understood at the first reading. This should be our target!

Always aim to use fewer words. Aim to use short sentences. When you have a choice, use a word that is commonly or widely understood, rather than a more 'scientific' one. Avoid repetition. As far as you can and the journal will allow, reduce the number of acronyms you use. These efforts at simplification will serve you well when you go on to write any scientific document or reports to your superiors or sponsors.

To be frank, science is often complicated, the techniques to study it are complicated, and the results too may require complex interpretation. Thus, writing or explaining it can be a real challenge. The good scientist is the one who can convey the important messages in a way that is crystal clear.

We acknowledge that this is doubly hard for those for whom English is not their first language. Nevertheless, there are some basic rules you can try to follow. First of all, slow down, stop and think. Think hard about what you are trying to say. Second, take part in as many discussions as possible with your colleagues about your work and do so regularly. Try to explain your research to your partner, friend, brother, sister or parents. If they look confused you will have to go away and try again. What you will inevitably find yourself doing is using more and more simple phrases and words to explain your work. This will reveal to you the need to explain things in simple terms if you are to get your message across. It's a necessity in science to remove ambiguity in what we say. In other words, you want to avoid a sentence having potentially different meanings. Here is an example: 'This shows that the treatment had less effect than expected'. Unfortunately the reader might not know which treatment and which effect is being referred to.

A useful tip when starting your career as a writer of scientific papers is to review successful (well cited) papers in your field. Note down helpful phrases that could be useful to you, such as 'these data suggest' (note that data is a plural). Then employ these phrases in your own work. Gradually this will build your confidence in using them and you will grow your own language to convey your science.

On occasion, the authors of this book feel that some of their fellow scientists resort to complex and opaque language out of fear of their not being seen as serious scientists, or simply because of the delusion that the language of science should, by definition, be impenetrable. Remember, to have maximum impact science needs to be understood by as many people as possible. Scientists need to develop the skill of explaining complex ideas and results in a manner that can be widely understood. Yes, this is not easy! A well known saying in literature is 'Easy reading is damn hard writing!'

Your motivation and understanding the needs of the reader

Your motivation should be to try and explain your findings to as wide an audience as possible. You should aspire to attract and draw in an audience beyond just the few specialists in your field. The natural tendency, particularly of young scientists, is to first and foremost placate the senior scientists and professors in their field. These individuals are feared like dangerous Komodo Dragons, irascible beasts who might tear off one of your legs if you contradict their cherished preconceptions. Consequently, a young scientist may be fearful and consider that the wisest strategy is to remain in the undergrowth, publishing modest but highly technical papers that do not contradict any established norm in their particular field. However, your ambition should be to interact with the many in your field and ideally those beyond it. Nor should you view your role in science as to always find and interpret the data in a manner that supports your supervisor's views or the established wisdom. Your loyalty in science can only be to the data; you go where the evidence directs you whether that is convenient or not.

Remember how you felt when you reviewed someone else's paper? You were tired and in a hurry; there seemed to be so many papers to read and so little time. Do you remember the papers that helped you the most? They were the ones which were clear and simple; they provided what you needed; in fact, they seemed to know what you were looking for! Whilst this is in part about the language and structure of the paper, the key is the attitude of the author and the efforts they have gone to in order to communicate clearly.

Your preparation

In a sense everything you do, at least in academic research, on a daily basis should be leading to a paper. Maybe there are some who swim around trying all sorts of things until, 'hey presto', they make a serendipitous discovery. But limited time and money usually don't give us such opportunity to wander far and wide in the hope of accidental discoveries. Prepare your hypotheses and carry out a series of experiments. Some of these will work and some won't. Keep asking yourself 'is this experimental data publishable? Do you have enough now to make a coherent story?' Judging when you have enough information to publish a paper can be difficult. There is a balance to be struck between rushing to publish the results from a single experiment, which produced surprising results, to waiting several years until you have repeated your observations many, many times. Typically, you are trying to marshal several different threads of evidence that together make a coherent story. Don't allow yourself to be distracted, doing lots of experiments which go in very different directions and, whilst 'interesting' (remember that dangerous word), are unlikely to contribute to your story and a paper. If you are a young scientist, many of these decisions, such as the course and direction of the science you are doing, are not in your hands. Nevertheless, you should not see yourself merely as a technical assistant. Instead, as far as you can, take control.

Practical aspects of getting started

We are assuming you or your leader have done a nice piece of research that is topical and mature enough for you to be able to say something. So set aside a decent chunk of time to write. Start by booking out two weeks (you may need more, but as you develop as a scientist you will require less). Do not try to write the paper alongside other activities, like working in the lab. Do not try to juggle several tasks during the day; you need to block out all distractions. Don't get drawn into your emails, helping others or chatting with colleagues. Stop doing administration! This may sound harsh and selfish but the only way to achieve the right state of mind is by totally blocking out the rest of the world. Get away from others, either in an isolated room or space or by using headphones to listen to music that will blank out competing sounds but not be so entertaining as to distract you.

Surround your workspace with print-outs of your best figures and tables. Let them inspire your story. Ideally, decide on one story and discard distracting information that does not add to your plan. Maybe consider a title early on. Have a chat with supervisors or your potential collaborators on the direction you want to take and the figures you want to use in the paper. Hopefully you get their support and it is time to start. It is essential that you alert and involve your potential co-authors from the very start. You certainly don't want them to rebel when you allow them belatedly to see your final draft!

Think of the structure of your paper

You will find some of our best scientists take a surprisingly short time to write a paper. This can seem inexplicable to many younger scientists who perspire for many months, if not years, in their struggle to complete one paper. The key to success is that a good scientist will have thought carefully about the structure of their paper before they start. They think carefully about what they should keep in and what they should discard in order to tell the story. So, for example, if you have generated 30 separate graphs from your research, you might choose only three for the paper. But these three graphs are the ones that are absolutely essential and decisive for the story. This careful thinking beforehand on what evidence to bring forward and what not to use allows for considerable economy of effort. It is annoying to spend a lot of effort describing a method and a graph which, on reflection, you don't need in the paper!

Then start typing as soon as possible, don't keep putting it off! At the beginning, the quality of what you write is not critical, you can edit later, the important thing is to get into the habit of writing. Your supervisors or managers can only help you if you hand over some text. Begin with headings and subheadings such as introduction, methods, results etc. Subheadings are particularly useful as signposts for the journey you and, ultimately, the reader will make. Then start chucking anything you already have lying around that might be suitable into these boxes. For example, you hopefully have figures and some methods already written from when you first did your experiments, so chuck them in. If you are not feeling so inspirational, maybe spend the day writing up those methods. At the end of each day you will have made measurable progress.

Personally, I enjoy writing the introduction as soon as possible in the process. Its purpose is to draw the reader into the importance of the issue your study is tackling. Your aim should be to review in an unbiased but candid way what we know already. But you do it from the point of view only of what that unknown reader far away needs to know. Then you end with a flourish as you outline the weaknesses of present knowledge and how you have chosen to address them. Rather as we mentioned earlier, when we suggested surrounding your desk with your graphs, here we recommend you surround your desk with piles of literature in folders which describe their content. This might be 'chemical structure', 'manufacture', 'waste products', 'market consumption' etc. Hopefully, these papers have been harvested by you over the course of the project, or even in the week before you started writing. You need to skim read them to, perhaps, remind yourself of their key information. In my case I put what I consider the most useful towards the top of each separate pile with my own scribbled comment in red biro on the major finding on the front page of each paper. In these modern and more enlightened times, you may have all of this in a virtual state on your computer, such as in a reference manager. But the key thing is you need to be able to write your introduction in a blast with your references close at hand. This process needs absolute focus. You are like a concert organist operating keys, pedals and stops simultaneously as you weave your introductory story. The longer you have spent thinking, discussing and reading about your topic beforehand, the quicker you can write that introduction.

- Decide what your main story will be.
- Stick with it, and be wary of side-tracks.
- Select the minimum number of figures and tables to support that story.
- Start writing with those figures and tables beside you!

The critical items

Getting the title right

It is hard to overstate the importance of your title. We estimate that when reviewing papers in our field, we first read the titles of perhaps 80–90%. Partly on the basis of the title, we then go on to read maybe 25% of the abstracts in our field. After reading the abstracts, we may only read the full papers of 5–10% of the work in our field. What this tells you is that your title is the most 'visible' part of your work. The obvious first task of your title is to be clear about the topic and work you carried out whilst remembering that brevity is extremely valuable. But you need to be offering the reader something that might be rewarding or attractive to them. So rather than saying '*The influence of cadmium, zinc and copper pollution on algal and macroinvertebrate populations on the Cherwell River one Sunday morning back in 1987*' you could improve this with '*The impact of metal pollution on the ecology of a UK river*'. Your abstract can then fill in the details on what metals you studied, in what river, on what wildlife and when. The important point is that your title hopefully will have made the reader interested enough to go on and actually read your abstract! So it is a good idea to try several titles and see which you and your co-authors prefer. Which manage to grab attention in the fewest words? Try to make as broad an appeal as you can.

A rewarding abstract

The next most likely part of your paper to be read is your abstract. This is usually the clinching factor on whether someone will actually go on and read the whole paper. You can consider the abstract as your shop window. You want to attract readers to come in and buy your work. The shop window does not display everything the shop possesses but it is representative. Most journals have a word limit for this section, often something in the region of 250–300. The basic task of the abstract is to be a very short description of what was done and to reveal your main findings. You can achieve this with introductory scene-setting in 1–2 sentences before describing the basics of your methodology.

Here you want to impress the reader with the scope of your studies, such as how many places you visited, how many samples you took, your use of a suitable analytical method, etc. This will provide the reader with some confidence on the robustness (repeatability) of your study. Reveal the most valuable part of your results and include some numbers. End by suggesting the implications of this research for the wider world. Thus, the abstract should be like a small polished crystal. It is small but attracts the eye. It tells the reader (a) that you know how to write; (b) you are not in a muddle (c) it will be an easy paper to read (d) the results will be clear and unambiguous, they will help him or her, and (d) you reached a clear conclusion. Writing a good abstract is a real skill; with practice you will get progressively better at doing so and also take pleasure in the challenge.

Key words

These are important because most indexing and searching for literature is done via keywords. So think carefully and choose good ones! Usually, general rather than specific keywords are best.

Maintaining a narrative thread (or the art of story-telling)

Although we like to think of ourselves as brutally efficient scientists working with robotic precision at any time of the day or night, we are not; we are human. That means, when we read someone's paper, we may be feeling tired or distracted by other conversations. Perhaps we are also feeling hungry and wondering what to cook for dinner tonight? These human characteristics are natural, but that means we authors have to work extra hard to keep our readers' attention. It is also often the case that when someone reads your paper it is with a particular purpose in mind; will it offer the key data or support they need to help them with their research? Both the human frailties and the limited requirements of the reader point to us having to make our papers simple and easy to digest. Please, please note that these weaknesses and the limited knowledge of our potential readers should not be viewed as their problem! It is our task to work extra hard to reach out to as many readers as we can.

The best way to keep your readers' attention is to write your paper as a form of story (just as we discussed when giving a presentation). Maybe you remember that, as a child, your anticipation grew as you wondered what would happen to those orphaned children left alone in the forest and you wanted to find out! This is a common human instinct we can connect with. So now you are not a cold-hearted and mechanistic scientist but instead a story teller, trying to put a smile of satisfaction on your reader's face! As a story teller we know we must select and focus on one good story, something the reader might care about. We do not wander around in an aimless fashion or distract with irrelevant details. This is called maintaining a narrative thread. We must bring our story to a close with a final rewarding conclusion. Although it may appear we are belittling the complex issues we all struggle with in science, a good story is an enormously helpful ingredient for a paper, ensuring it is well received; we are humans after all.

Please use lots of subtitles!

A way to help and structure your story is through the use of frequent subtitles. These help to turn your possibly long story into a series of digestible chunks. There are few things more daunting than reaching the discussion section of someone's paper only to discover eight pages of heavy, densely worded text subdivided into only a few huge paragraphs. Brave souls might plunge in, but soon they may be lost, confused or simply bored. The information they need may be in there, but where? So you must subdivide your discussion into a logical sequence of topics with their own subheadings. The structure, as shown in your subheadings, might go something like this:

- The major findings.
- Other possible interpretations.
- Comparisons with existing literature.
- Limitations of your study.

Super clear figures

In theory, you could publish your paper just as a sequence of figures. These should be capable of telling your story in their own right. But for your figures to successfully communicate will require effort on your part in once again considering the needs and limitations of your reader. On the one hand you might have the noble desire to be transparent and show everything but, on the other, you must avoid distraction and confusion. You are there to guide the reader through the complexities of the scientific bramble bush to the fruit of your conclusions at the end. Thus, you will have two major choices to make:

- Which figures to use?
- How to display your data to best advantage?

At the end of a long and often complex project you will be in the position of having several figures you could show. All of them have merit and represent miniature stories in their own right. However, as we discussed above, we strongly recommend that you focus on one major story in your paper. You may wish to run some parallel minor stories alongside, or note additional implications of your work, but you must try to avoid distracting the reader from your main thread. Therefore, you should select only the figures which effectively drive forward the main story of your paper. Fortunately, journals allow you to add a 'Supplementary Information' section where you can place your extra data and figures. So the second issue is how to present your data. You must recognise that there are generally several approaches you could use. This is actually an opportunity for you to play and have fun! Different approaches may come to you over several days or perhaps be stimulated by discussions with a colleague. Show and discuss your figures with your co-authors or colleagues at tea-break. Do they convince? Do people immediately grasp what is being shown? Papers and their figures are all about the art of persuasion. Thus, the reader will be particularly interested to see your controls and also other co-variables. They will want to see the variance around your mean or median. If you are looking at changes in several locations to the same driver and variable, then use the same scale. In an X–Y graph you may truncate the Y-axis so that it no longer starts at zero, but be careful, as this may be seen as an attempt to mislead readers as to the magnitude of change. Never forget you are a communicator! If you place several graphs in one figure and then sprinkle those with abstract letters (e.g. treatment c) and acronyms, you are making them very hard to follow. Your reader may end up missing the point and go back to thinking about what to cook for dinner! The basics are:

- Each figure tells only one story.
- It does so extremely clearly.
- It does so honestly.

Conclusions section

Most journals have a conclusions section as a final subtitle, but not all, although all readers expect that you will provide a summary of your main findings. There is great variety in how authors approach this section. Some leave no conclusions at all; they have taken you into the dark dense wood of their data and expect you to find your own way out. Some go to the opposite extreme of exaggerating the importance of their findings way beyond what their evidence can sustain. More commonly, authors provide a long laundry list of conclusions, sometimes returning to a meandering discussion, and so leaving the reader bemused as to what of real importance has been found?

We certainly do recommend you provide conclusions, but make this section as short as you possibly can. Ideally, you begin by returning to the original hypotheses you proposed at the start of the paper and consider whether these have been falsified. Keep to only a few points for the reader to take away, because he or she will not be able to remember several. If they cannot remember your main finding(s), it is less likely they will cite your paper in their future papers. In your final sentences you must plot a path between avoiding exaggeration and not underestimating what you have achieved.

Dealing with reviews

In the happy eventuality that the journal will consider publishing your manuscript, as long as you constructively respond to the reviews of your paper, you will now have a new job to tackle. Quite often you will be facing several pages of apparently annoying and suspicious reviewers' comments, and it will feel as if you have a mountain to climb. But almost always, by responding to reviewers' comments you will end up with a better paper. You might think that they have indeed found some very serious flaws which shake your confidence. It might even seem as if you have to go back to the lab and do many more experiments or re-analyse thousands of data points. But don't despair!

The first thing we recommend is to construct a table and place every single reviewer comment in sequence in boxes in the first column. Give each reviewer's comment a number. Now re-read your paper and the comments over a period of a few days. This thinking period will allow you time to consider how best to respond (and calm your emotions). It may well turn out that many comments are variations on a similar theme and can be responded to in a similar way. Think how your existing data, and perhaps that in the supplementary information, can be marshalled to respond to the comment. Now, in your table, provide your response in the box next to the numbered comment. In this box you can acknowledge the point being made, or at least your interpretation of the criticism, and then guide the reviewer and editor to the information and data which will support your case. Where you strongly disagree with a comment or viewpoint, you can defend your opinion in your response to the editor. The journal does not insist you must always change your paper to suit a reviewer. Then, in the next column, tell the editor in what line and page you made a change in your manuscript. Try to avoid inserting large quantities of new text in your manuscript in response to the reviewers. Do the minimum necessary to keep the clarity and direction of your paper intact. On the other hand, you may be asked to reduce your text. Whilst this may at first seem very hard to do, it is possible, and usually leads to a better paper for others to read. When editors see you have made a sincere and comprehensive effort to respond, they are likely to accept your paper for publication.

Problems for the science writer

Fear of confronting that blank sheet of paper

If you have been putting off writing your paper for days, weeks or months, and when no further escape or prevarication is possible, being finally confronted with that blank sheet of paper (or computer screen) is scary. As you look at the many papers by other authors around your desk you may say to yourself, maybe I can't do this, perhaps I am not up to this. Converting your work into a scientific paper is the point when you arrive and announce yourself to the world as a scientist. Unlike your other outputs, be they reports to sponsors, conference papers, oral presentations or posters, here you will be judged by other scientists on whether your work is worthy of publication. In other words, you are running the risk of being judged and rejected, with all the damage that can cause to your self-esteem.

To get round this problem of 'freezing', let us get back to basics. Science functions and moves forward not by word of mouth, or on the basis of a brilliant analysis over a few beers in the pub. It does so by the sharing of work and ideas in scientific journals. You may have been working long hours in your lab in Malaysia doing intricate experiments that evoke the admiration of your co-workers, but another scientist sitting in a different country will not know about you or your successes! The rest of the worldwide science community will not know about them either until you have published that paper. So scientific journal papers are how we, as scientists, communicate with one another across continents and time. We won't learn about you and your efforts if you remain mute. So please come and join in the conversation! Remember that, with the internet, your paper will become available to everyone almost instantaneously, and that it will be there in perpetuity. You will never be forgotten, however small your contribution.

The curse of perfection

Many scientists verge on being perfectionists. Whilst this is laudable, it can prevent them from ever publishing. They are telling themselves and their colleagues that their data are not yet sufficient or of high enough quality. They say 'better to wait till I do some more experiments, it should only take another year – or two'. There are a number of problems with this strategy:

- Timeliness – the big need for the information could be now. Several years later, many others, possibly with much greater resources than you, will have published their results and your work will then have little impact.

- You may already have made much more of a breakthrough than you realise. Immersed as you are in the topic, you may be worrying that experiments 5 and 6 didn't quite go to plan, whilst perhaps forgetting that your previous experiments 1 to 4 have made great steps!

- What do we mean by perfection? We have yet to read, or indeed write, the perfect paper ourselves even after

30 plus years and hundreds of attempts! There are good and less good papers. A paper is a contribution to an ongoing dialogue.

So please don't dwell on perfection, or the apparent perfection of your supervisor or other scientists out there. Make the most of what you have got; don't underestimate what you have already achieved; the important thing is to get started.

The most common mistake

When people share their manuscripts with us to get advice, they nearly all share the same common mistake. Their manuscripts are **far too complex!** They assume the reader comes to the paper with a vast existing knowledge of the subject, and then they plunge them into confusing vortices of mystifying complexity. Please, please, DO EVERYTHING YOU CAN TO SIMPLIFY!

Final thoughts on paper writing

The more you practise, the better you will get! Your role is not to tell people all you know, but instead to provide them with what they need to know. You must acknowledge you are a story teller and so your duty is to attract the reader and draw them inexorably to a (preferably single) conclusion. All the time you must try to make life easier for your reader.

There is a theory that a good paper simply means one that has been accepted by a 'high impact factor' journal.

The impact factor is the average number of citations earned by papers in that journal two years after publication. However, there remain many papers in high impact journals which have low numbers of citations and, similarly, you can find highly cited papers in less well thought of journals. The internet has brought with it the democratic judgement of the worldwide scientific community, and this is what really counts. Thus, scientists now compare themselves based on the total citations their papers have received (minus self-citations) and on a variation of this named the 'H-factor', after its introduction by an academic called Hirsch (Hirsch, 2005). Thus, if you have an H-factor of 26, that means 26 of your papers have 26 or more citations.

Although we as the writers may entertain fond hopes that by reading this chapter you will overnight become an excellent writer of scientific papers, we accept that this may not happen. So you need to find a key motivator or motivation moment in your life that will spur you on to take up the quest of continually aiming to improve your writing. This flame may be ignited through seeking out and learning at the feet of scientists whose papers are highly cited, or through bad experiences of finding that people do not understand your work. You must find what lights and maintains this flame of desire to write well and communicate your science. But, and this might surprise you, there is pleasure to be gained by crafting a clear sentence and well-focused paragraphs. Stand back and admire them. Whisper it quietly, you could almost call it a work of art!

Checklist for Chapter 10: Writing a (good) scientific paper

1   First of all, review your data and identify what is unique and special about your paper. This will now be your focus. Stick to this main focus/finding!

2   Think through the structure of the paper before you start writing. Only use material that will really contribute to the story.

3   Structure your paper with lots of subheadings, don't be shy!

4   Have you gone through every single sentence in the paper to make sure it is as short as possible (a sentence should be no more than 1–1.5 lines in A4 12 point)? Organise carefully to avoid repetition.

5   Have you taken every step to make sure the paper is as short as possible (to maintain the focus)? Fewer words are ALWAYS better than more!

6   Avoid complex phrases and keep to simple words.

7   Avoid using personal terms such as 'our' or 'we'. Science is supposed to be impersonal and objective.

8   Use acronyms sparingly and ensure they are properly explained at first use.

9   Make figure legends full and complete and explain as much as possible (date samples taken, replication, etc.).

10 Do not make the figure the subject of the sentence! Discuss your results and end the sentence with the figure/table in brackets. For example 'the highest concentrations of Zn were in the Yellow river (Fig. 6)', rather than 'Figure 6 shows . . .'

11 Do not repeat in the text long lists of data which you have already presented in the figures and tables. Results and discussion should be a limited summary of the main findings. The data in the figures/tables do not need to be repeated in the text.

12 Do not use several significant figures without good reason! A value of 7.6 is better than 7.5894! This is particularly true when model estimations are given, because here there will be large uncertainty.

13 Ensure that the amount of discussion is proportional to the importance of the topic. Do not distract the reader with long discussions on aspects that are trivial compared to the main focus of the paper.

14 Maintain that narrative thread! You must keep the readers' attention so that they can follow your story. Don't run several different stories or mess things up with information that distracts from your storyline. If you try to put too much stuff into a bag, it will break!

15 Use emphasis carefully. Don't say 'this clearly shows' as that implies certainty and hints at arrogance. Try instead to use 'this strongly suggests' or 'this indicates' which, whilst revealing your conviction, still leaves a space for uncertainty.

16 Work hard to squeeze out any ambiguity. Try to make each sentence stand alone and not require the preceding sentence to make sense. Put the paper down and re-read after a week. Does it still make sense?

17 The conclusions section should be as brief as possible, a paragraph of no more than 1/3 of a page. Don't re-open the discussion. If permitted, bullet points are very useful.

Reference

Hirsch, J.E. 2005. An index to quantify an individual's scientific research output. *Proceedings of the National Academy of Sciences of the United States of America* 102, 16569–16572.

CHAPTER 11

Writing grant proposals

Sources of funding

Research costs money; often an awful lot of money. That money often comes directly or indirectly from government. The former usually means a government department, whereas the latter source includes organisations such as the National Institutes of Health (NIH) in the US or the Research Councils in the UK, which receive their budgets from government and spend them on supporting research. Funding can also come from charities, where medical charities dominate, or from industry. Each potential source of support for your research will have specific goals, so to be successful you need to focus your research on these goals. For example, a cancer charity will fund only research relevant to cancer, so if your research is on another disease, look elsewhere for funding. Although that example is obvious, in other cases it is not so easy to know whether or not your intended research is relevant to the funding organisation you have in mind. In the case of industry, it is usually easy to know if your research might be of interest to it. Each industry tends to be highly focused on one topic: the pharmaceutical industry will not be interested in funding research on improving the technology behind mobile phones and the telecommunication companies will not be interested in supporting research on new pharmaceuticals. The difficulties arise with curiosity-driven research – sometimes called 'blue skies' research – when you have a novel, fundamental idea that you would like to investigate. This novel idea may have no obvious practical use, in which case you want to do research for research's sake. The richer countries of the world often have funding organisations that will support curiosity-driven research, although it is probably true to say that support for this type of research has steadily declined during the careers of the authors, and been replaced by more applied research that addresses specific problems requiring solutions. This change has, to a large extent, occurred because public funding organisations have come under increasing pressure to demonstrate that the research they support has societal impact. In other words, public funding organisations need to be able to demonstrate that the research they funded has been useful to society. This desire to demonstrate relevance and impact has led to major changes in the information requested from scientists applying for grants, as we discuss below.

In summary, there are usually many potential sources of funding for your research. Time spent investigating which one, or ones, would be most interested in your research is time extremely well spent. You do not want to waste a lot of time writing a grant proposal to an organisation that has little or no interest in the research you want to do.

Seek help and advice from the funding organisation

A very sensible way to ensure you have identified an appropriate funding organisation is to engage with it before you write one word! In the same way that you do not want to waste time writing a research proposal that has little or no chance of being supported (i.e. funded), funding organisations do not want to waste their time considering grant applications that stand little or no chance of being funded. A very wise move is to speak to potential funding organisation before doing anything else. You will almost certainly find them very helpful. They will tell you what funding schemes they have at the time and whether or not your research interests fit well with their current interests. They will tell you the factors that are important to them. For example, do they want to fund small, short-term projects or longer-term, larger projects; whether or not applications can include international partners; how projects are reviewed, and in what timescale, etc., etc. (see also 'Keep to the rules' below). Being better informed can only increase your chances of success; it can sometimes do so very significantly. It is perfectly legitimate to contact a funding organisation to seek advice: do not be afraid to do so.

Consider small schemes to get you started

Sometimes the chances of success of an application for a relatively small amount of money are greater than the chance of a large grant application getting funded. Particularly if you are at a relatively early stage in your independent scientific career, it is probably wise not to be over-ambitious and apply for a large grant. Instead, consider applying for a reasonably small sum of money. That small amount of money may not be large enough for you to employ someone to do your research, but it will enable you to begin your independent scientific career. It might, for example, enable you to buy a piece of equipment you need, or pay for you to spend some time in another laboratory that has facilities you would like to use. The psychological boost you will receive from obtaining even a small grant can be very significant, so do not underestimate small successes. Very good research can often be done with quite small amounts of money and other resources. The results from small pilot studies can provide you with vital ammunition if you choose to apply for much larger funding subsequently. That way you can demonstrate to the grant reviewers that you have the expertise and preliminary results that offer the promise of success at a bigger scale.

Keep to the rules

All funding organisations, with the possible exception of industry, will have their own, unique application forms and large amounts of guidance on what should be included in each section of the form. Follow that guidance closely. Give the funding organisation all the information it requires to assess your proposal, even if you think that some of what is requested is of little relevance to your proposal.

One of the more difficult sections of an application can be where you need to say what impact the proposed research will have. More and more funding organisations around the world now ask for this information, often in a section entitled 'Impact Statement'. Your research is unlikely to change the world, so do not overstate its likely impact. But do not be too humble and cautious either. If you think that your research will have significant impact, say so, explaining clearly why you think it will.

Your institution is likely to have its own 'rules' about grant applications. These days, applications often have to be submitted by your organisation rather than by you directly. If this is the case, then engage with the appropriate part of your institution as early as possible. It is likely that there will be people in your institution whose job it is to help scientists write grant proposals. Those people may have much more experience in preparing proposals than you do (see 'Getting help' below); they will certainly know more about the bureaucracy involved in submitting a grant proposal than you do. For example, they will know how to accurately cost a proposal: you may not. You have no choice except to work with these people, so try to do so as a team; after all, all of you have one aim – to submit the best proposal you can, while making sure that you follow all the rules. You do not want to fail because you did not provide some information that was requested.

Getting help

As suggested above, a lot of help can be provided by your institution. Accept it all. Besides helping with all the information that you, as a scientist, might consider as bureaucracy, help with the science is probably also available. Do not be afraid to discuss your research idea with colleagues: they will usually be very helpful and supportive. Even if they are not experts in your field, you will find that explaining your idea to a non-expert can be a very good way to ensure that it is very clear in your own mind. If he or she cannot understand it, or appreciate why it could be important, then your thoughts are probably not as well-developed as they need to be. Consider writing a short summary, perhaps in one paragraph, of your proposal, then giving this to two or three colleagues you respect to obtain their opinions on it. Another strategy that can be very beneficial is to give a seminar on your proposal to your colleagues. The audience is likely to be very mixed, consisting of both early-stage and established scientists. Some will probably know quite a lot about your field of interest, and hence understand your research idea easily, whereas others in the audience will be 'non-experts'. That mixture can often be ideal for testing out your idea before you write even one word. Try to be receptive to suggestions for improvements and even criticisms. There is no sense in seeking advice from others if you then ignore it. Instead, use it constructively to improve your proposal.

Finally, as your proposal nears its final draft, many institutions now have an internal review process in which other members of staff review your proposal, doing so as if they were reviewing it for the funding agency you intend to submit it to. Often the proposal cannot be submitted by your institution until it has undergone this internal review. Try to respond positively to any critical comments. If you think that a reviewer has not understood something in your proposal, do not conclude 'he/she does not have the knowledge to understand my proposal', but instead think 'how can I make that bit clearer so that other scientists will understand it?' If your colleagues cannot understand something in your proposal it is likely that external reviewers also will fail to understand it. If that is the case, then your chances of getting your proposal funded are extremely slim. Do not ignore advice that has been provided to help you.

Hypotheses and aims

The most important part of any proposal is the science: what do you intend to do, how will you do it, and why is it important to do it? The 'what' question is the key: exactly what do you want to do? Put another way, what is the scientific question or issue you want to tackle? A good way to convey what you want to do is to provide a hypothesis: what falsifiable hypothesis will you test (see Chapter 3 on the use of hypotheses in research)? You might then have two or three (not ten or more!) specific aims that, once completed, will have thoroughly tested your hypothesis. Both authors of this book have read many grant applications which were not clear about exactly what scientific question was to be tackled. Proposals can often be full of the fine detail to be applied to the many aims and objectives, yet not state clearly how these will help test a hypothesis.

After stating very clearly – and hopefully very simply – the scientific issue you will address with your proposal, preferably by testing a hypothesis, you will need to provide supporting evidence that your proposal is plausible. That is, can you actually do what you say you want to do? It can impress some reviewers and funding organisations if you propose using new, innovative techniques, perhaps techniques that now enable a question to be tackled that could not be addressed until that technique was developed.

But try not to let exciting new techniques dominate your proposal to the extent that quite what hypothesis you want to address gets lost.

Do I need collaborators?

It can often strengthen a grant proposal significantly if you include a good collaborator. By 'good' we mean someone who brings skills to the proposal that you do not possess. He or she may well have unique expertise in a key area of science relevant to your proposal. They might also be a more experienced scientist than you, and hence have had some success in winning research funding, possibly even from the organisation you intend to submit your proposal to. They should be willing to work with you in preparing your joint proposal; if they are going to be a good collaborator, they will be happy – possibly even enthusiastic – to put effort into helping you write the proposal. You may think that by having a collaborator you are giving away your scientific idea, as well as giving away some of the money if you are successful. But half a grant is very much better than no grant, so if you think that your proposal will be strengthened by having a collaborator, bring one on board.

Grant applications for large projects requiring large amounts of funding can involve contributions from many scientists, sometimes based in many different countries. Here in this book we are focussing on younger scientists trying to start their independent scientific careers by obtaining their first external grant, and hence we will not discuss the strategy required to obtain a large, multi-centred grant.

The wow factor!

Given the wide range of competitors, it is worth thinking about anything special or unique about your bid. Why now? Why hasn't it been done before? What is new and exciting? Do you have a new piece of technology, one that has never been applied in this field before, or perhaps a large dataset that could throw new light on a problem? What great, exciting deliverables will you offer at the feet of a grateful funding organisation?

Allow enough time

Writing a good grant application cannot be done in a day! It probably cannot be done in a few weeks. It usually takes months to write a grant, although not every minute of every day has to be spent writing it (you will have other things to do). So start early: as early as possible. Let others, especially the administrators whose jobs are to help scientists write grant applications, know your intentions even before you begin to write. The earlier in the process they can provide their contributions, the more efficient the process. Likewise, if you decide to seek the opinions of some of your colleagues, or decide to give a presentation on your idea, do so earlier rather than later. If your organisation requires internal review of an application before it can be submitted, you will need to have the finished application available a few weeks before your submission deadline. Many scientists try to do everything at the last minute. Hence they are writing major sections of their proposals only a few days before the submission deadline. This will inevitably lead to a poorly written proposal, which is likely to stand little chance of being funded. Submitting a proposal that you know could, and should, have been better is very unwise and a very inefficient use of your time. Very experienced scientists might be able to write a proposal in a few weeks, but you are unlikely to be able to, so start early. Frequently, grant proposals involve more than one organisation and individual collaborating together in an application. Rather than relying on email communication between partners, you must organise one or more face to face meetings to ensure you all understand what is required.

Success or failure?

Unfortunately, failure is more likely than success. This is simply because the competition for research funding is intense, and hence the success rate is low. A 20% success rate is normal, and the rate can be significantly lower. Thus an average scientist will need to write five to ten grant applications to succeed with one! Such poor odds could lead anyone to conclude that they have better things to do with their valuable time than write grant applications that stand little chance of being successful. And indeed, many scientists do decide that they will not try. This is probably the main reason why many young scientists never become established, independent scientists. Many try once or twice, become demoralised when their applications fail, and give up. It takes a lot of determination and confidence to keep trying but keep trying, you must if your research cannot be done unless it is funded externally, as is the case for many scientists, especially those in universities. And always remember that if you do not try, then your success rate will definitely be zero!

Often you will receive feedback on your application, sometimes before the funding decisions have been made and sometimes afterwards. In the former case, you are likely to be given the opportunity to respond to that feedback, which is likely to consist of reviews of your proposal by anonymous reviewers. Do respond and do so constructively, in the same manner as you would respond to reviewers' comments on a paper you are trying to get published (see Chapter 10: Writing a paper). Of course, you cannot change your original proposal in any major way – that would require a new proposal – but nevertheless you may be able to suggest changes that increase your chances of being funded. Any increase is worthwhile! Even if it seems obvious from the reviews of your proposal that it probably stands no chance of being funded this time, make sure that you learn from the feedback, so that your next proposal is better. The same applies if the feedback comes at the same time, or after, the funding decisions have been made: use criticism to improve.

Join grant reviewing or moderating panels yourself!

The best way to gain a perspective on how a reviewer examines a grant proposal is to volunteer to review grants yourself for a funding organisation. You might also try to join the higher tier, which is the moderating panel. This group is usually there to sift through the reviews and then do the final whittling down to which projects to fund. Typically, this experience will show you that a good grant proposal can never be too clear or explicit about its aims/hypotheses. Clarity is also vital in explaining how the proposal will drive science forward. This clarity is needed because the panel will in truth only discuss your proposal for a short period; they have a lot to get through. If they don't understand (and they won't necessarily be experts in your field) what you want to do and why in a short space of time, you will fail.

Checklist for Chapter 11: Writing grant proposals

---

1   Do your research on funding organisations and carefully analyse what a call for proposals may be requesting.

2   Enlist as much help as you can!

3   Having collaborators with different skills can add to the quality of the science but you must meet them and ensure you understand each other.

4   Clarity will be vital in your writing.

5   State early on what hypotheses you are testing.

6   Find a way to bring in a 'wow factor' with your offer.

7   Get started early; time will run out fast!

8   Learn what makes a good proposal by becoming a reviewer yourself!

CHAPTER 12

How to cope with rejection

You may not realise it, but it is likely that you have already had to cope with rejection. You may well have been rejected even before you became a scientist! This is because you probably had to apply for a number of PhD positions or other types of junior scientific posts before you were successful and obtained one. You must have managed to cope with those rejections and ultimately been successful in obtaining a scientific position, otherwise you would not be reading this book. Almost all scientists, even very good, very experienced ones, have to learn how to cope with rejection. Their ideas might be rejected by their supervisors; their papers might be rejected by reviewers and editors; and later on in their careers their applications for funding might be rejected by funding agencies and they themselves might be rejected for promotion or a new job they applied for. How scientists respond to rejection will determine how their careers develop, and how good a scientist they ultimately become.

Your ideas are 'rejected' by your supervisor

Let us assume that you already have a scientific position – perhaps you are a PhD student – and that your supervisor has encouraged you to plan an experiment. You will be excited that at last you are about to do some real research. You will want to demonstrate to your supervisor that you are very knowledgeable in your field of research. To do this you design an extremely complex experiment that you are certain will, once completed, resolve every important problem in your field in one go. You proudly show this experimental design to your supervisor, who after a very brief consideration of it will probably reject it! They will say that it is far too complicated. Exactly this situation occurred to one of the authors of this book, and more than once, too. The supervisor would spend no more than one minute looking at the first attempt at a 'wonderful' design, then look up and say "simplify, simplify, simplify". That was the end of the discussion. I went away feeling both rejected and dejected. It was a very humbling experience. During the following few days I modified my experimental design to simplify it, then presented it a second time to my supervisor. It was again rejected! At least this time it received a few minutes of consideration from my supervisor before it was rejected, almost always because it needed to be simplified further! Often, but not always, my design was accepted at the third attempt. Only now, three decades later, do I understand how much I learnt from those very humbling meetings with my supervisor. I now realise that he was probably the best designer of an experiment that I have ever met; he knew how to design an experiment to maximise the chances of it providing data that could be interpreted easily and clearly. And he passed those skills on to me through rejection of my over-complex designs. I learnt that rejection is probably necessary during the development of a scientist, because only through rejection can we improve as scientists.

Your paper is rejected

Your next rejection is likely to occur when you submit your first paper to a journal, hoping (and expecting) that it will be accepted for publication in that journal. By this stage in your development as a scientist you will probably be feeling reasonably confident: you know your field of research well, you have obtained some interesting results, and you want to see your name in print. Then others will know that you have arrived as a real scientist! After much drafting and redrafting, you and your supervisor have produced a paper; one that you both feel is a good paper. You send the paper to the journal of your choice (see Chapter 10), and a few weeks later receive the response. You nervously open the email and soon come across the following words from the editor, "I regret to inform you that the manuscript cannot be accepted for publication". Your lovely paper has been rejected. Your first reaction on reading those words will be a wave of emotion and despair not unlike the rejection of a love letter! You will feel that YOU personally have been rejected: the response from the editor seems to be saying that you are a poor scientist who does poor research, research that does not meet the standards required for publication. On the day of the bad news it is best to keep busy with other tasks and not respond, just give yourself time for the emotions to calm down.

When your feelings are a little calmer you will need to let your co-authors know that the paper has been rejected. This will be embarrassing, so you may decide to do it via email rather than face-to-face. However, doing so is often surprisingly positive, because now the rejection is shared with others and hence you will feel that it is not you, and you alone, that has been rejected. Doing so will initiate discussions with your co-authors. You will discuss the opinions of the reviewers of your paper: are they fair and balanced? Doing this is almost always very encouraging and positive. It is extremely rare for anyone to blame you or anyone else for the rejection. Instead, your co-authors usually work together very effectively to decide how best to proceed. The options range from giving up on the paper, if you conclude that the reviewers are right, through to resubmitting the paper to another journal. You need to decide whether or not you have faith in the paper. Usually the answer is yes. If you do, and hence you decide to submit it to another journal, the wise scientist redrafts the paper, taking into account the criticisms, comments and suggestions of the reviewers of the initial copy. Almost always reviewers have things to say that, if addressed constructively rather than dismissed as 'unfair', 'wrong', or 'ill-informed', improve a paper, often a lot. So rise to the challenge and improve your paper to increase its chances of being accepted second time round. Doing so is by far the best response to rejection of a paper.

You will need to get used to having papers rejected for publication because it is likely to happen to you quite a lot. Do not assume that once you become an experienced, established scientist with a reasonable track record of publications your papers will always be accepted, rather than rejected. The writer of this chapter has published over 250 papers in scientific journals and is, he hopes, considered a good scientist, yet his last paper was rejected by the journal it was first submitted to. And the paper before that one was also rejected first time round. This happens to over half of all the papers he submits to journals – and these are not being submitted to prestigious journals such as *Nature* and *Science* that have very high rejection rates. His most highly cited paper was rejected by not one, but three, journals before, eventually, being accepted by the fourth. That was a very humbling experience, but also one that aided my development as a scientist because it taught me a lot. My book co-author can still recall that his first major paper from his PhD was rejected by the reviewer with the words 'this paper should not be accepted by this journal or any other!' The message is that having your papers rejected is common and hence you must learn to cope with their rejection if you are ever to become an independent scientist. Remember that you are not alone; you will struggle to find a scientist who has not had to cope with rejection of one or more of their papers. If he or she survived that rejection, you can too.

Your application for a job is unsuccessful

Now let us assume that your first position in science, which is often as a PhD student, has gone reasonably well and that you have enjoyed your initial foray into scientific research, but now it is coming to an end and you need to find your next position. This might be a job in industry or government or as a post-doctoral fellow in academia. You may be lucky and be given a position (usually temporary) where you currently are, but it is more likely that you will have to seek a position elsewhere. Doing so will require you to apply for positions that become available. You will probably not be successful with the first application you send, so prepare yourself for more rejections!

It is difficult to advise someone on how best to cope with rejection of a job application because, unlike with papers, there is rarely any feedback to explain why your application was not successful. You would like to know, to help you improve future applications, but usually no reasons are given. There are, of course, many reasons why you did not get the position you applied for, some of which you can influence but others you cannot. For example, if there is a strong internal candidate (which you are unlikely to know), your chances will be low, or if the position requires someone with specialist skills that you do not have, but which were not stated in the job advertisement. If you want to become a scientist, you have no option but to keep applying for scientific positions until, hopefully, you succeed and are offered one. Your strategy for coping with this rejection needs to be (1) do not give up – keep trying – and (2) improve your Curriculum Vitae (CV). There are two ways that you can improve your CV and application. One is the obvious one: publish more papers, give presentations at conferences and use these for networking, and do some teaching, especially if you are seeking a university position that will involve a significant amount of teaching, as many do. We will discuss this more in Chapter 13 on professional development. All of these will strengthen your CV and hence hopefully increase your chances of being invited for interview, rather than having your application rejected, the next time round. The second way is less obvious: improve the presentation of your application and CV. I have appointed a lot of junior scientists, and hence read many hundreds (possibly thousands) of applications. I am always surprised, and disappointed, at how badly many people present themselves in their applications. Seek advice and help on how to write a good application. A lot of advice is available online. Think carefully about the characteristics required to do the job you are applying for, and make sure that you emphasise what you have to offer that aligns well with those requirements. But do not exaggerate: be honest. And make sure that your CV looks good visually, so that it catches the eye of the person or persons sorting through all the many applications received in order to decide who to invite for interview. There is often intense competition for scientific positions, especially those that offer long-term, or even permanent, contracts, so unsuccessful applications will be much more common than successful ones. Keep this in mind when you are applying for jobs. Just because your application was unsuccessful does not mean that you are a poor scientist, nor does it mean that you will be unsuccessful subsequently. Learn to shrug off the disappointment of a failed application for a job. See it as a challenge to do better next time.

Your grant application is rejected

Let us now assume that you have been successful in obtaining a scientific position. You are now on the way to becoming an independent scientist, a position you have wanted to reach for the last few years. This will probably mean that you will try to obtain external funding for your research. In Chapter 11 we discuss how to write grant proposals in order to maximise the chances that they will be successful and win funding. However, in most countries the chances of a grant proposal being funded are low: a 20% success rate is typical and 10% is not unusual. Put another way, only one or two out of every ten applications get funded! Hence your chances of having your application rejected are very much higher than they are of it being successful! If you are going to build an independent scientific career, it is clear that you are going to have to learn to cope with rejection, probably with many rejections.

Can I survive all these rejections?

Yes, you can. We did, as have many other scientists. Probably all scientists, at every stage of their careers, will experience rejection. It is a common bond between us and, over a drink, every scientist can recall some particularly harsh rejections over the course of their careers. Of course, this situation is no different from many other professions. If you are going to become a successful scientist, it is necessary to learn how to cope with rejection. You will need to be confident (but not too confident!) in your own abilities. Do not take rejection too personally. Rejection of a paper for publication, or a failed job application, does not mean that you are a failure as a scientist. Both authors of this book have experienced many rejections, and still do, yet have become relatively successful scientists. You can too.

Checklist for Chapter 12: How to cope with rejection

1   Rejection in science is a normal part of life and in part reflects the 'self-correcting' nature of science. Every scientist suffers!

2   When it happens don't lash out or over-react! Let the dust settle for a day or two.

3   Having ideas rejected or criticised by your supervisor is better than being knocked down publicly later on. The subsequent refinements could lead to greater success in your work.

4   Having your paper rejected happens to everyone, to good and bad scientists. But you can bounce back with submission to another journal.

5   Console yourself that many theories we now accept and use were initially rejected!

6   Despite the temptation, don't assume your grant or job application was turned down purely because of some problem in your character.

7   You will survive!

CHAPTER 13

Interacting with the science community through social media

Accessing information and interacting with other scientists

When I began my scientific career in the mid-1970s the only way to keep up with the literature was to go to the university's library and browse through the relevant journals, and the only way to interact with scientists with similar interests was to attend conferences. Both strategies had major limitations. The main problem with the library was that it might not have the journals most relevant to my research area. Only a few universities and research organisations could afford to subscribe to a full range of scientific journals. It was much more likely that your library had some of the journals you wanted it to, but not all of them. Hence it was almost impossible to know if you were unaware of some of the most important literature in your field. The main problems with conferences were that you could attend only one or two each year, due to the costs of doing so, and that it was quite likely that the particular scientists you hoped to meet and talk to were not attending the same conference.

The situation is very different now: keeping up with the literature and interacting with scientists with similar interests to you, wherever they are in the world, is very much easier now. It was the invention of the internet that dramatically improved both access to the literature and interaction between scientists. The internet enabled the development of databases such as the Web of Science and Google Scholar, which not only contain the majority of the published scientific literature (with the exception of much literature published in languages other than English), but allow that literature to be searched in different ways, such as by subject or author. Now, most scientists rarely, if ever, go to a library! Instead, they browse the literature on their computers while at their desks, or on their tablets or mobile phones wherever they happen to be, and at any time of the day, too. Keeping up with the literature is now very easy, even though there is very much more literature published each year now than there was in the past.

In contrast to the fundamental changes in how scientific literature is accessed, the format of conferences and what one can gain from attending them has not changed much during my scientific career. Scientists still present the results of their research through oral presentations and posters, and these provide opportunities to meet and talk with other scientists who are also attending.

The development of the internet has led to a wide range of platforms being established that enable people, including scientists young and old, to interact with each other, to get to know one another, to promote themselves and their research, to keep up with developments in their field of interest, and to learn about job opportunities.

One such platform is **ResearchGate**. This tends to be used mainly by established scientists, who place copies of their papers there, enabling other scientists to easily access those papers. You can also put information about your current projects on your profile. Discussion between scientists is possible – someone might ask you about one of your papers or seek your advice – although this function is relatively rarely used. The platform provides very regular information to you about how many people are reading your papers and where those people are located. It can be rewarding, and interesting, to know that some of your publications are being read by scientists located at an institution on the other side of the world; but is that information nothing more than a small ego boost? How does it help you improve as a scientist? Many scientists consider ResearchGate to be primarily somewhere you deposit your extended CV online, and hence they use it more as a self-promotion platform than as a way of providing information that will enable young scientists to improve themselves.

LinkedIn is another online platform that can be considered primarily a way of promoting yourself. It is very popular not only with scientists but all professionals. Essentially your profile on LinkedIn is a mini-CV that you are using to 'sell yourself' to others; it summarises your experience and achievements. The key basic facts are there, not the detail (no copies of papers, for example). LinkedIn does have a messaging system, an internal email system, which can be useful.

Many scientists, especially younger ones, now use social media both to become informed about scientific advances in their field of interests, and also more generally to interact with other scientists with similar interests to their own. When this book was written, the two most widely used platforms were **Facebook** and **Twitter**, both of which have hundreds of millions, if not billions, of users. Of the two, Twitter is considered much more useful to scientists than Facebook. Many laboratories now have a presence on Facebook, but really that is only for self-promotion. A few scientists do use Facebook for science, but most people use it to talk to their friends. You can have your Facebook page open, and hence allow anyone to engage with you, but you will probably soon find that you are learning little, if anything, of scientific relevance to you. It may well be best to use Facebook for your social life rather than for science.

Twitter is a completely different 'beast' compared to Facebook or any other social media platform. Users can use only a very limited number of words and pictures in their message or 'tweet'. Twitter is super-fast with scientific news, and hence it allows you to keep up with developments in your field very easily. Some scientists consider use of Twitter as the number one way of keeping up with developments in their fields of interest, a 'third eye' if you like. To these scientists, use of Twitter is a required part of their daily lives and they might spend an hour, or more, on Twitter each day, both posting and reading. News comes, and goes, on Twitter very quickly. Nothing seems to last more than two days, but in that time a piece of news can circle the world and be read by thousands of people. Twitter explains science in a very concise, user-friendly way. It is a very easy way to gain the attention of a complete stranger who is nevertheless interested in what you have said, and it means you can find out who you are reaching. Some of these people will tweet only occasionally, but some do so very regularly. You can easily monitor these regular tweeters, as they are called. Twitter also provides plenty of metrics that inform you about how many people liked your post, read it, forwarded it (re-tweeted it), etc.

Besides posting your own scientific comments – possibly about an interesting result you have just obtained, or a paper of yours that has just been published, or a comment on a paper by someone else – and reading the tweets of others, there is an aspect of Twitter that is much less obvious. Twitter creates a personality for you. People learn much more about you than you might think when they read your tweets. You will, unintentionally, be cultivating an image of yourself. Be aware – very aware – of this. Use it to your advantage, not disadvantage. Make sure that you carefully manage the picture you create of yourself. Remember that many people can know you through Twitter; many more than through any other mechanism, such as face-to-face meetings. You can use Twitter to get to know people and learn whether or not you have things in common, for example opinions on scientific issues of interest to you both. Then, if you do meet face-to-face at a conference, you are not scientific strangers, nervously feeling your way with each other. Instead, you are already friends, and hence you can get straight to the position you would like to; a discussion of your mutual scientific interests. Put another way, Twitter can help you decide who you would like to meet. It can be a science dating App! It enriches your scientific career by helping you decide who you like scientifically and personally. However, it takes time on Twitter to build the number of contacts you need for a reasonable network, so be patient; do not assume that your first tweet, however interesting to you, will be read by hundreds or thousands of fellow users of Twitter. But, slowly and steadily, you can build a network of people who are interested in your research, as you are in their research. Once you have a network, you can flex it, including removing people that you no longer want to hear from because you are not learning anything worthwhile from them.

Twitter can be used by supervisors to alert students to papers that they would benefit from reading and, by using the appropriate hashtag, you will be able to learn in 'real time' about what is being presented at conferences, whether or not you are present at the conference, and you can participate in that discussion if you want to. You may also learn about potential new junior positions, such as available PhD studentships or post-doctoral fellowships, via Twitter, because senior scientists and organisations can use Twitter as a good way of disseminating very widely an advert for a position.

Another platform that is becoming popular with young scientists is **Instagram**. Instagram has significant and currently underutilised potential as a vehicle for science communication. The visual nature of the platform, combined with its large and diverse user base, makes Instagram an efficient way to engage with the public about science, thereby increasing science literacy and promoting trust in science. It can also be used to explore career options.

Personal blogs do not seem to have become as useful as many scientists thought they might be. Some scientists even consider them 'a flash in the pan'. If you run your own blog, you need to keep it up to date, to ensure that it is always interesting, as well as demonstrating that you are organised. You can alert other scientists to your blog via Twitter, if you want to.

Many universities and other types of organisation involved in scientific research now provide courses in the use of social media by young scientists. Find out if your organisation offers such courses and, if it does, take whatever is offered. Doing so could constitute a very well spent half a day or day, at most.

The limitations of social media

Like all advances in technology, social media can be both positive and negative for your development as a scientist. Use it in a smart way and stay focused on what you are supposed to be doing: learning to become a good scientist. Do not spend time chatting to your friends on social media when you should be focused on science, however tempting it is to do so (see Chapter 8: Time management). It is very easy to waste a great deal of time on social media.

Keep in the forefront of your mind that none of the social media platforms will provide you with a full picture of any scientific issue relevant to you. One or two hundred words in a tweet cannot replace a scientific paper. Remember also that someone very active on social media might be biased and primarily interested in promoting themselves: be very aware of this possibility.

Everything changes

More than any other chapter in this book, we needed to obtain advice from other, younger scientists in order to be able to write this chapter about the use of social media by today's scientists. That was because things change: how today's young scientists access the literature and hence keep up to date with advances in their fields is different from when we were young scientists, spending many hours each week in a library, searching laboriously through journals, looking for articles of interest to us. And it is inevitable that things will continue to change, and this may mean that the content of this chapter is soon out-of-date. Nevertheless the basic philosophy underpinning the chapter – use technology wisely and remain focused on your science – will remain highly relevant.

Checklist for Chapter 13: Interacting with the science community through social media

---

1. There are opportunities here to grow knowledge of your work.

2. **ResearchGate** is a convenient and accessible shop window for your work and projects.

3. **LinkedIn** provides a more straightforward professional CV and networking opportunity.

4. **Twitter** allows you to track science issues of the moment and also get to know the characters of others you might work or collaborate with in future (and they get to know you).

5. Judge carefully the amount of time you commit to social media.

6. To help grow your knowledge and inform others, be prepared to adapt to new technology and platforms as they come along.

CHAPTER 14

When things are not going well

Unfortunately, there is a very wide spectrum of things that can go wrong in your research. Pressures include failed experiments, declining relationships with work colleagues and anxieties about 'the need to succeed'. People react differently but, particularly when you are young, a series of setbacks can begin to affect your self-confidence if not your health.

When someone begins a research career, he or she will look forward enthusiastically to the next few years of research. They will feel very positive about the future. Perhaps it is inevitable that this high level of enthusiasm cannot last; very few research projects go well from start to finish. It is much more likely that at some stage your research will not go well. There will be periods when you make little, if any, progress. At times you may even think that your research is going backwards. Failing experiments and self-doubt, together with working alone, can become threats to your well-being. The authors of this book are scientists, not therapists, and hence here we are not attempting to diagnose any mental health problems you may experience, or suggest suitable treatments for those problems. However, as supervisors and mentors to many young research scientists, we have learnt quite a lot about the anxieties many students experience at some time during their studies, and we have gained experience on how students cope when their research is not going well.

My own PhD 'failed'

My PhD did not go as planned. I aimed to purify a hormone involved in the reproduction of fish, then establish a technique to measure blood concentrations of that hormone during the reproductive cycle of the fish. My laboratory had a lot of experience in the techniques I would use, and I had very good supervisors. Yet I failed. I never managed to purify the hormone and thus I had no pure hormone to use in establishing an assay capable of measuring blood concentrations of the hormone. A few minor parts of my PhD did provide some positive results, but the main objective of my PhD was not achieved. Thus, early on in my scientific career, I learnt about failure and how I could cope with it. I did not realise it at the time, but this experience was to prove extremely valuable to me later on, both in helping me deal with my own subsequent failures and disappointments (i.e. papers rejected and grant applications not funded) and in helping me supervise my own students when their research was not going well. The experience is not uncommon; many scientific projects 'fail'. For example, we know that many scientists spend their whole lives in the pharmaceutical industry working on compounds that in the end fail to become drugs taken by patients. But these scientists have not themselves 'failed;' they have carried out rigorous science from which we learn more about what different compounds can and cannot do, so that we can have better medicines tomorrow.

You are not alone

It is easy to constantly feel inferior to your fellow researchers. You may think that they come up with all the good ideas, and that their experiments always work and hence provide good results. You may think that you are the only one who is struggling, whereas everyone else is sailing smoothly along. Yet you would be wrong. Survey after survey of junior researchers has reported that anxiety and depression levels in postgraduate students are high. It is often said that they are unacceptably high. It is unclear if they are higher than they were in the past, although there is much anecdotal evidence that rates of anxiety and depression are rising year-by-year. Perhaps the pressures on young scientists are greater today than they were in the past. We are all more visible in the internet age. This makes the pressure to succeed more of a public struggle. As I discuss below, realising that you are not alone if you are struggling with your research is the first, and possibly most important, step to take in dealing with your problems.

Anxiety about your research, and the more serious condition of clinical depression, are not conditions confined to academically weaker students. In our experience, academically strong, apparently confident students are just as likely to become anxious about their research, to the extent that they cannot function effectively.

Do not try to work yourself out of trouble

When things are not going well, many of us respond by working harder. Our strategy is to work harder, doing more experiments, in the expectation that eventually things will start to improve. Surely if we put in enough effort we will eventually be rewarded with success? Working late into the evening and then through the weekend can be isolating and damaging to your home life. Often we do not confide in others, especially our supervisors, when we are striving hard to get an experiment to work. We want to meet him or her when we are feeling good because we have some interesting results to discuss, not when we are feeling demoralised and have only failure to talk about. But whereas the temptation is to work harder and harder, pushing yourself too hard can have very negative repercussions. Try to know your limits; don't push yourself too hard. Doing so rarely, if ever, gets you to where you want to be.

There is more to life than science!

To keep up and perhaps stay ahead it is tempting, when you leave the lab, to spend your evenings and weekends reading about science, keeping up with the latest scientific topics on social media and watching science videos on TV or the internet. The danger is that this continuous absorption prevents your brain from getting a rest. It may perhaps contribute to an existing anxiety you have that somehow everyone else is succeeding in science except you. Do please refresh yourself in the evenings and weekends with friends and family. Indulge yourself in an absorbing hobby, something that takes you completely away from your science career and research.

Seek help and advice

It is very common to hear or read something like 'I knew I was struggling, but I did not ask for help. I thought I could deal with my problems'. The instinct of most people is to battle through any problem on their own, hoping that they can resolve it without the need for help. Many people feel that telling others about their problems would demonstrate that they were weak individuals. But the reality is quite different. It is not a weakness to seek help if your research is not going well despite all your best efforts, particularly if this is influencing your mental state. It might initially take some courage to admit to others that you are having problems, but it is definitely the wise thing to do. And there are many people you can turn to for help and advice. It may surprise you to read this, but it is our experience that everyone you confide in will be sympathetic and helpful, often because they too have had problems with their research and the associated stress. There may well be other research students and post-docs in your research group, or in other research groups located nearby, whom you could talk to. Some of these young scientists will likely already be your friends, and hence they should be both very understanding of your difficulties and very helpful in trying to find solutions to your scientific problems.

Researchers we have discussed these feelings of failure with say the first step towards turning things around was to unburden themselves to others. Some organisations have welfare officers who have a role in supporting young researchers. They can help you put things in perspective. They will very probably reveal that you are far from alone in having had such doubts and unhappiness.

Two, or more, heads are better than one when it comes to solving problems. You should also talk to your supervisor(s). You might be apprehensive about doing so, but it is the right thing to do. I expect that you will find your supervisor very helpful and supportive. Good supervisors realise that their students need most support from them when things are not going well. Besides caring about your welfare, supervisors get most from their students when their confidence returns and they are happy. Hence supervisors have more than one strong incentive to help their students deal with their problems, both practical and emotional.

Despite what I have said above, remember that supervisors are not trained to deal with serious mental health problems; they are scientists, not professional counsellors. If you are experiencing serious mental health issues – and the incidence of these appears to be rising sharply in young people, including young scientists – then it may be wise to seek professional help. Most large organisations, such as universities and research institutes, provide counselling services where you can talk to trained people in complete confidence. Do not be afraid to seek their help if you think that you could benefit from it.

Take time off work if you need to

As stated above, it is tempting to deal with a scientific problem by putting in more and more effort, and longer and longer hours, in the belief that eventually things will get better. However, by doing so you are likely to keep repeating the same mistake. If you are really struggling, to the extent that you have lost most or all of your enthusiasm for your research, then it may be wise to have a break from work. If you think that this is what you need to do, then make sure you discuss this option with your supervisor first. Do not simply 'disappear', leaving your supervisor with no idea where you are, or when you will return to work. Try to think of their position as well as your own. A break from work for a few weeks, or even a few months, can be very beneficial; it can give you time to think logically about your problem and how best to tackle it, and it can help improve your mental state. There is no shame in taking such a break to 'reset' and refresh your mental state. The important thing is to recognise that downward spiral. You can then return with a clearer mind and a broader perspective about your project and indeed your life.

Serious depression

The incidence of serious (clinical) depression in young scientists appears to be rising rapidly in many countries. The reasons for this situation are unclear, but probably involve young people feeling that they are expected to succeed and yet have relatively little influence and control over their lives. If you become clinically depressed, you definitely need professional help. Your fellow students and supervisor(s) are not trained to deal with depression, and should not try to do so. Seek professional help and then act on the advice you receive. Clinically depressed scientists should not be attempting to do research; they need to recover fully first before returning to their studies, if that is what they decide to do.

Summary

The falsification of hypotheses (sometimes described as the trampling of beautiful theories by ugly facts) and simple experimental failure are part and parcel of the research experience. In other words, failure of experiments or great theories being disproved is the experience of all scientists, including the very best. This mixture of falsification and experimental failure, to say nothing of grant application failures and paper rejections, will continue throughout any good scientist's career. But younger scientists may consider such events as equating to a personal failure. This is particularly the case when they are combined with time pressure and a feeling of isolation. All parties, from junior scientists to senior managers, need to be aware that the territory of science brings with it certain pressures, real and imagined, that can at times seriously impair performance and health. By acknowledging this we can help one another.

Checklist for Chapter 14: When things are not going well

1   Problems, mishaps, experimental failures and rejection are common to the research experience of all scientists.

2   If you are feeling anxiety and self-doubt about your research, you will not be alone. This is not a failure or weakness in you.

3   Don't assume that working longer and longer hours will solve all your research problems.

4   If stresses are really getting to you, then open up to friends, welfare officers and your supervisor(s) about it. It is essential your supervisor knows because he or she will be part of the solution.

5   Maintain a healthy and fulfilling life outside science. Avoid 24/7 immersion!

6   Be prepared to take a complete break if you need to. Refresh your brain.

7   Seek professional help and advice about depression early, don't fight it entirely on your own.

CHAPTER 15

How to be a better supervisor

This book is intended to help people beginning their scientific careers to progress to becoming good scientists. Thus this chapter, and the next one (Chapter 16), might seem out of place here and irrelevant to young scientists. But they are not. There are, in fact, a number of reasons why young scientists should be interested in the content of these two chapters. An obvious one is that you may, in the future, become a supervisor yourself. If you do, it is very likely that you will regularly ask yourself questions like 'Am I supervising my students well?' and 'How should I supervise my students?' In this chapter we offer some of our thoughts on supervision based on the experience we have gained while being supervisors ourselves, as well as what we have learnt from observing how other scientists supervise their students. But not only could these two chapters help you in the future; they could help you now. It is inevitable that you will frequently think about your supervisor(s), and will ask yourself questions such as 'What level of support should I expect from my supervisor?' and, more generally 'Am I being well supervised?' These two chapters (Chapters 15 and 16) aim to help you view the research student:supervisor relationship from the point of view of the supervisor.

Every young scientist who has successfully completed a PhD will know how important the relationship with their supervisor(s) was. Yet it may come as a surprise to learn that few institutions provide much, if any, training to supervisors. The author of this chapter has supervised over 40 PhD students, many coming from countries other than the UK. Yet his university provided him with no training on supervision and even now, nearly 40 years after his first PhD student began her studies, it provides only very limited training. Much of the training provided nowadays is concerned more with protecting the reputation of an institution (often a university) from complaints, and even litigation from disgruntled students, than it is with helpful advice to a supervisor on how best to supervise a young researcher. Common sense says that a new manager – for that is what a supervisor is – should be routinely offered the training that will help them in their new role. As science becomes progressively more international, with young scientists in particular moving from country to country to do PhDs and post-doctoral research, supervisors can find themselves supervising students coming from very different cultures to their own, which adds further challenges to supervision.

It is extremely important for supervisors to realise that PhD students are not technicians. Nowadays a supervisor is much more likely to spend most of his or her time in an office on a computer than in a laboratory or in the field doing his or her own research. In this situation there can be a temptation for supervisors to use their research students as their technicians. If established scientists require help to do their research, then they should hire a technician. Supervising a technician is very different from supervising a PhD student. With a PhD you are preparing someone to take decisions in science on their own, to think for themselves. A PhD should be a training in research: someone beginning a PhD will have little, if any, experience of doing research, yet by the end of their PhD they should be well on the way to becoming an independent scientist, even if, as is likely, they have not yet reached the final stage in their development as a scientist. The job of the supervisor is to support and inspire PhD students, so that they can make this transition. Take the job very seriously: the future of a young scientist is in your hands.

When a student starts, they literally do not know what to expect. The standard that appears to be expected, be it from written work, talks or posters, can appear to them to be dauntingly high. They will be concerned not only about the judgement you might make on their work, but also about the opinions of other professors in the department, or of those staring back at them at their first conference. Granted, you will probably feel that their first offerings are far from perfect, but remember that they are at the beginning of their project and your student will be feeling vulnerable and anxious. So always try to find something positive in their work and tell them you are encouraged by their progress. Shortly before they go up to give their first talk at a conference, tell them how impressed you were with their previous effort or practice talk!

Getting the relationship right

To do a good job as a supervisor it is necessary to develop a close relationship with a PhD student. Although in some countries PhD students have a team of advisors, it is almost always the case that one person is the main supervisor and he or she does the majority of the supervision. The supervisor and student need to establish a professional relationship built on each respecting the other. This can lead to quite a close relationship developing; try not to let it get too close. You will need to meet regularly. It is difficult to be precise about exactly how often supervisors and students should meet, but it should certainly be at least once a month, and probably significantly more frequently. Encourage your student to take notes of the meeting. It is a good idea for the student to keep a hard-back notebook dedicated to this purpose. At the end of the meeting go over the major points which have been discussed with the student once more. Where there is more than one supervisor and you all have a joint meeting with the student, the student should take minutes of what was agreed and circulate these afterwards. However, do not constantly pester your students every day; they will never develop confidence in themselves as scientists if you do. It is likely that there will be periods when you need to meet quite frequently, and other periods when less frequent contact is fine. There is a very strong temptation to meet a student very frequently when their research is going well and they are producing nice results – after all, as a supervisor you should be very interested in those results – whereas if a student is going through a quiet phase, you may feel that there is little point in meeting them. However, this strategy is wrong! Students need to meet their supervisors most frequently when things are not going well; this is when their confidence will be at its lowest, and when your support and encouragement can be very positive. Supporting your students in difficult times is much more important, although also often more difficult, than supporting them when their research is going well.

Honesty and realism

The relationship between supervisor and student needs to be a trusting one. A high level of integrity is vital in science (see Chapter 5), and hence it is imperative that students are open and honest to their supervisors. So, early on in a student's PhD, a supervisor should discuss the issue of integrity with the student. The supervisor must feel certain that what he or she is being told by a student is the full story, including details of any errors the student may have made, or anything that the student knows went wrong. Although, ideally, a supervisor should see all the raw data a student produces and perhaps even retain copies of them, in reality my experience suggest that this situation is very rare and may never be achieved. Most often the supervisor sees only a summary of the data, a table or graph, for example. Hence the supervisor relies absolutely on the honesty of the student. This will not be achievable if the student feels under pressure from the supervisor to deliver 'good' results, those results that he or she thinks their supervisor expects and wants. So supervisors should never say to their students things such as 'I need those results by Friday' or 'I am giving a talk at a major conference next month and want to include some new, exciting results from your research'. If you put a student under pressure to deliver results, do not blame your student if the results you are provided with turn out not to be repeatable!

Learning to let go

The supervisor's approach to supervising a student should slowly but steadily change during a PhD. Initially the student will be very inexperienced, know little about the topic they are studying, and have little or no appropriate practical skills. They will require quite a lot of support initially, but they will learn fast and as they do so they will require progressively less support from their supervisor. The supervisor's job becomes one of helping the student gain the confidence required to (eventually) become an independent scientist. Put another way, the supervisor needs to slowly but steadily 'let go' of the student as they grow and mature as a scientist. This can be achieved by, for example, letting a student decide what research to do next. Perhaps they have an idea that they would like to explore; if they do, and their idea fits well with the topic of their PhD, encourage them to explore their idea. Doing so will demonstrate to them that their supervisor has confidence in them, and it will be a significant step in their journey towards becoming an independent scientist. Remember that you are trying to train a young scientist who, at the end of their PhD, will probably leave you and your supervision, and who will need to have enough confidence in their abilities as a future independent scientist to move on to the next steps in their career.

Dealing with problems

Of course both supervisors and students hope that no significant problems will arise during the three to five year supervision period. but sometimes they do. How they are dealt with will have major consequences for both the supervisor and the student. The aim should be a positive outcome, not a negative one.

A wide variety of different problems can occur. Perhaps the most common one is that the student does not make the amount of progress that the supervisor expected. In such a situation, the supervisor needs to discuss the situation with the student, to try and find out why progress has been slow. There are many possible reasons, ranging from scientific ones (e.g. the equipment keeps breaking down) to personal ones (e.g. the student may have health problems, financial problems, or relationship problems). When you talk to the student, do not be confrontational; this is completely the wrong approach and one that is very unlikely to get to the bottom of the problem. Try to be reasonable, understanding and, if appropriate, sympathetic. Learning the cause of a problem is a necessary first step to resolving it. But do not get too involved in issues outside of your knowledge and expertise. For example, if the student has a health problem – mental or physical – encourage him or her to seek medical advice and support. Remember that you are the student's academic supervisor, not their doctor! If a student is reluctant to tell you what his or her underlying problem is, there may well be other, specialised, people at your institute who may be able to help the student.

For example, nearly all universities have a free counselling service available to students (see Chapter 14: When things aren't going well). Many also have a graduate school where appropriate expertise may be available. Students may also feel more comfortable speaking to someone who is not their supervisor rather than talking directly to their supervisor. Let them. That is why many institutes have regular appraisal meetings at which students discuss their progress with a small group of academics which does not include their main supervisor. Such meetings can often tease out problems and suggest solutions to them, when a one-to-one meeting between student and supervisor cannot. Supervisors should not feel threatened by these meetings; they are not intended to undermine the supervisor.

Supervisors, rather than students, can sometimes be the cause of problems. I have already mentioned that putting students under pressure to produce 'good' results is both unprofessional and unethical. But supervisors can cause other problems; for example, if they move to another institution or get promoted within their own institution, their students are likely to be affected, possibly adversely. If a supervisor moves to another institution, a decision has to be made as to whether or not her or his research students should move with them. Many factors will play a role in that decision. Full and open discussions should take place with the student(s) and each student's wish should be honoured. Forcing a student to relocate will almost certainly end in a mess, so if the student, for whatever reason, would prefer to stay where they are, then alternative supervisory arrangements need to be made. This does not mean that the original supervisor cannot continue to play a role in the supervision of the student. If a supervisor is promoted, or changes his or her role, he or she may have less time available to supervise students. If you find yourself in such a situation, be honest about it and, if necessary, in discussion with the student(s), change the supervisory arrangements so that you play less of a role while someone else takes on a greater one. Always do what is best for the student. That will almost always be the best for you as well. I have known many occasions when academics have tried to continue supervising research students after promotion (to Head of Department or to a senior administrative role, for example). I cannot think of a single case where the student has not been disadvantaged, sometimes very significantly. If you do not have the time – and it requires a lot of time – to supervise a student well, then do not try to do so. When taking on new duties and responsibilities it is always wise to relinquish some of those you had previously.

Supervising a post-doctoral researcher

Obtaining a secure, possibly permanent, scientific position is very competitive, and it is usually necessary for a young scientist to complete a few years as a post-doctoral researcher before their CV is good enough for them to stand a reasonable chance of succeeding in obtaining that coveted scientific position. Post-doctoral researchers often do much of the research in a research group. As they have chosen to remain in research after completing their PhDs, they are very often characterised by having considerable passion for, and dedication to, their research. Yet the postdoctoral phase of a young scientist's career is usually dominated by short-term (one to three year) contracts, and hence job insecurity. In many countries many post-docs will not find a position in academia – there just are not enough jobs available for the present high number of young scientists with PhDs – which is a situation that both supervisors and their organisations need to keep in mind when supervising and supporting them. A number of research-intensive institutes have recently established postdoc centres in order to improve support and professional development for postdoctoral researchers.

Keeping these facts in mind, a supervisor should adopt a quite different approach when supervising a postdoctoral researcher than for a PhD student. A postdoctoral researcher may not quite yet be the 'complete researcher', but they have already been a scientist for a number of years. They bring with them knowledge and experience that you may not have. Thus it may be more effective for the supervisor and post-doctoral researcher to consider themselves as a team, rather than the relationship being one in which the supervisor is 'the boss'. In fact you may rely on them to help with the supervision of your PhD students. Try to give a postdoctoral researcher as much freedom as they want. Some will be quite happy to look to you for advice and guidance, but others – the more independent ones – will want to play a significant role in deciding what research they should do, and how it should be done. Do not be afraid to let them, by giving them the flexibility they need to develop into fully independent scientists. You may well be surprised how advantageous it is to you to do so. Remember that if you combine your knowledge and experiences, you are likely to achieve more than if you assume that you alone should dictate the direction of the research: two minds (you and the post-doc) will almost always be better than one.

Checklist for Chapter 15: How to be a better supervisor

1　Your attitude is critical to the success and development of your students.

2　Remember that those training for a PhD are not 'hired hands'.

3　Make deliberate plans to meet those you supervise regularly (between once and four times a month); don't leave it to chance.

4　Work on building a relationship of trust and honesty.

5　Recognise they will need you most when things don't go to plan.

6　When progress is much less than expected, do not be confrontational, try to get to the bottom of problems and, if needed, summon additional resources to help.

7　You want your chicks to be able to fly the nest, so build their confidence and encourage their independence.

8　When complex issues arise, doing what is best for your student will also be best for you.

CHAPTER 16

Wider aspects of science management

There are many who would argue that managing scientists is close to an impossibility, something akin to 'herding cats'. On the whole scientists are fiercely independent and driven by a passion for their subject. By and large the Universities and Research Institutes have tolerated, if not encouraged, this independence of mind and spirit. Nevertheless, some level of management is essential to ensure the money spent on research projects produces tangible results, to prevent duplication of effort, and provide a fair provision of resources. This is particularly needed in large organisations. However, in the West, many academic organisations seem to be experiencing a creeping managerialism, where managers become more remote, the burdens of electronic administration increase, and direction is lost. But what are the essentials of good management of scientific staff?

The first thing to note, surprising as it may seem, is that you, as a manager, 'set the weather'. Your approach to your staff and your relationship to science will have a profound influence on their performance.

Management models

There can be many managerial roles in a scientific organisation, from finance to health and safety, but here we are concentrating on how to manage and motivate scientific staff. Firstly, you need to consider whether you are there only to manage in the administrative sense, to provide scientific leadership, or to nurture your staff.

- The Manager
- The Leader

The basic or limited view of a manager is someone who simply ensures everyone completes their timesheet correctly and who gets on the case of those who don't. This might keep the wheels ticking over, but little love will be lost in either direction, and both parties will feel frustration. Hopefully, you have ambitions for your staff/ department/organisation to generate ever more impressive scientific outputs. A good leader will inspire others to go the extra mile by choice. They will fight like a tiger to support their staff and constantly look for ways to help encourage and inspire them. To use a military analogy, a manager only gets the troops into the front line trench at the right time, whilst a leader gets them to go over the top and charge the enemy. However, an alternative way to get results might simply be to nurture the budding talent you have available in a pastoral sense. Gentle support can also be very successful in improving morale and performance.

Start by establishing your core principles

We would suggest:

- Make excellent science your priority and do everything you can to facilitate this.

- Encourage, enthuse and support staff in writing grant proposals. But this should be proportionate to the chance of winning and the funding available.

- Support collaboration amongst scientists of different disciplines and expertise. In particular, look out for opportunities to make links with the very best external scientists and thank staff who appear to be reaching out in this way.

- Identify key external policy-makers relevant to the science of your group and ensure they get to hear of the great science being carried out and its potential value.

- Recognise that your staff are your greatest resource. The success of the Department depends on their efforts. Keep abreast of their publication outputs and grants won, so that you can send a note of congratulations.

- Do what you can to limit, and ideally, reduce the burden of non-scientific administration placed on your scientists.

- Make time to regularly meet and chat face to face with your staff to encourage them, but pay particular attention to junior members of staff and their needs.

Do let your staff know that these (above or similar) are your principles, maybe even pin them to your door! After that, you must lead by example and embody the values you want to see in others.

Recruiting well

As the most difficult problems in organisations are invariably personnel related, the obvious advice must be to 'recruit good people'. If you have good people, no matter how poor your management and accounting systems are, things will go well. If you have bad people, no matter how good your management and checking systems are, things will go badly. It may be that you are looking for someone to become a leader and drive things forward. Unfortunately, it is surprisingly difficult, particularly at an interview, to identify who would make a good member of staff. By good, we mean a hard-working, constructive, pleasant person and excellent scientist. We have been misled in the past by those who appeared to have a high degree of technical/background knowledge, or who were very confident individuals at interview, yet who turned out to be bad appointments. For a potential leader we want someone who has energy and desire. Energy and drive cannot be taught! When trying to recruit leaders for the future, it is far better for the organisation that you select candidates who might appear to threaten you with their knowledge and determination than select those who simply pose no threat.

What are the clues to finding good people in the interview process?

- Those who prepared very thoughtful and honest covering letters which revealed that they have done their homework on you and the department.

- Those who reveal in their CV, letter or interview that they have overcome adversity either in their personal or academic life (thus, they have evidence of dealing with problems in the past and have the drive to succeed).

- Those who have suggestions for how they can take the department or project further forward in novel directions that you had not considered. This is in contrast to those who provide CVs telling you what the work will do for them!

- Those that take full advantage of the opportunity, when offered, of asking you questions. Some advocate that a considerable chunk of the interview should be set aside for this. Once again this will show the people who have done their homework on you, and indeed want to map out their future. It can reveal the hunger and desire of a really serious candidate.

- If you are aiming to recruit a research scientist at a more senior level, i.e. post-doc or above, then the key indicator will be publications and an ability to win grants, with the most valuable indicators being consistency in publication and first-authorship. From reading this book you will now have a good idea of what to look for in these publications!

Some of our most valuable staff did not have great CVs, nor did they stand out as interview performers, but we found more valuable qualities buried within them!

Project management

Often the best and most interesting science is done by multi-disciplinary teams working to address complex problems. These teams can include members from different countries, sometimes even in different continents. However, a successful outcome for science teams is never a foregone conclusion. The project manager has to grapple with three main issues: quality, time and budget. Which should he or she prioritise? As the purchaser of this book, you will know that we believe scientific quality must always be paramount, but a good manager will need to control time and budget carefully to ensure success. In big projects, further delegation will be needed to identify task leaders. It is vital to know who is accountable. So what are the best guarantors of project management success?

- Where possible, choose collaborators who have a track record of being able to deliver and, where necessary, compromise.
- Have a clear and simple plan that is practical.
- Never underestimate the potential for confusion and misunderstanding! So ensure your requests are very clear (not just to you).
- Have regular face to face meetings with the team, and make sure these have an agenda and minutes with agreed actions. The frequency is of course a matter of judgement and depends on the project duration and distance between partners. Having your partners present their results and compare these against agreed deliverables will help to concentrate minds!
- Ensure the actions you've agreed are not forgotten.

Holding meetings

There are many good reasons for holding a meeting, but choosing how many to have calls for a high degree of finesse. Too few and direction may be lost, too many and you lose productivity, plus staff could become resentful as their work load grows. Chairmanship of meetings is a real art and perhaps an undervalued skill. You have to combine

supporting wide-ranging debate and democratic inclusive-
ness with a dictatorial ruthless in accomplishing decisions
and keeping to time. This starts with setting a realistic
agenda and ends with accurate minutes and agreed actions,
with indications of who will carry out those actions.

Maintaining trust

Increasingly in the modern world we bring in a range of
checks and balances to reassure us that things, or more
specifically people, are doing what they are supposed to.
This can be justified by reference to the rare cases where
people have been selfish or foolish. But computer-based
management systems which drill down in ever increasing
detail into how staff are using their time can have a negative effect, because the essential and continual message
appears to be 'we don't trust you'. This feeling is neither
inspiring nor motivating. Sadly, it will encourage some to
'game the system', whereby they appear to fill their diaries
with extraordinary numbers of activities and meetings of
little value. If you have recruited good people (see above),
give them as much freedom as possible for them to deliver
the goods. They will be empowered by your trust.

Don't confuse process with purpose

The purpose of large science organisations should be to
produce great science and the scientists of tomorrow. But
large organisations increasingly use a whole series of pro-
tocols and software support systems for staff to carry out a

range of activities. These range from staff appraisals, health and safety documents, and purchasing, to funding proposals and time management; the list goes on. These various and completely different software systems are upgraded or changed with depressing frequency. Each will require training that needs to be continually refreshed. None of them seems to work well or be in any way intuitive, and they all require extraordinary input from you, the scientist. There is a danger that these 'beasts' require continual feeding and that this need starts to become a higher priority than the science itself. Those who fail to keep pace can be subject to both private and public criticism. This gives a signal that the organisation appears to favour and reward those who are best at keeping up with the administrative tasks, whilst doing good science takes a back seat! And yet all of these processes were meant to make an organisation run in a smoother and more efficient way, so that it could focus on its key purpose – doing great science!

Support initiatives from others in your department

There will be times when someone in your department comes forward with an initiative which you had not considered. It may not be completely thought through and the timing may be far from ideal. But actually, the important thing here is not so much the idea, but rather the person. If someone is coming forward with suggestions for the greater good, they should be encouraged. You are then publicly supporting those whose instincts are to improve the department.

If your first instinct is to pour cold water over the idea, not only will it depress and demotivate the individual, but it also sends a discouraging signal to other staff.

Praising staff can bring extraordinary benefits

For reasons that aren't entirely clear, it seems very much easier and more natural to criticise rather than congratulate others on their performance. Praising good performance, even over small matters, seems a lot more 'unnatural' to the manager. This should not be so. In fact, praising staff, whether they be junior or senior, will cause their morale to sky-rocket for months or even years on end. They are now more likely to go the extra mile for you, and less likely to be off work sick. This is particularly important where it is difficult, if not impossible, to raise someone's salary. Many organisations provide the opportunity to review in a written form a member of staff's annual performance. Rather than this being taken as another bland administrative duty, the chance should be grasped to give genuine, tailored and heartfelt thanks and congratulations where appropriate. But if you use the same form of words for every member of staff, they will not value it.

Give opportunities to individuals

You should have the development of your staff at heart. Thus, it is wise to identify opportunities for staff to take on and stretch themselves a little. The more they take on and succeed, the more their careers can progress and the more valuable to the organisation they become. They can feel more fulfilled as they succeed at different tasks, and you learn more of their potential. However, you may also discover that others are actually comfortable in their current role and would feel alarmed at change. But if they recognise you are looking out for their interests, this will still be a good outcome.

If you have to criticise . . .

This should always be done in private, not by email and never ever by email copied to other parties. The individual must be given the opportunity to explain their actions. The whole issue might well be a misunderstanding. Note that a sin of omission is much less serious than a sin of commission (where a deliberate decision has been made to do something wrong). Don't forget that the person will still be a member of your staff tomorrow. You must offer the carrot, not just the stick.

Defend your own staff in public

There may be times when others, perhaps much more senior than yourself, criticise someone in your department. This criticism may or may not be well-founded. As the manager of this department, your first instinct should be to stand by your own staff, or at least offer mitigating factors on their behalf. You can find out the true facts of the case later. Immediately and publicly joining in the criticism will have a chilling effect on all your staff when they learn you did not support them. They will no longer see you as a leader to respect, since they now know that when the chips are down, you have no real loyalty to them.

Time to re-organise?

Both research organisations and universities are 're-structured' with depressing regularity. Each new director/ manager entertains hopes that their new structure will in some way stimulate and energise the scientists. Perhaps a re-branding is actually genuinely needed to improve the marketing of the organisation, or departments need to re-align to be more in tune with new generations of customers or students? But if you wish to carry out re-organisations, do provide your staff with evidence to support this decision. Why is it necessary? It may be the right choice, but be aware that considerable disruption (particularly for other senior scientists) will occur and also that anxiety will ensue. Some scientists may fear that reorganisation means they no longer fit this new structure, and in some way they are being singled out to be disposed of.

Use evidence to support your decisions

Use evidence to support your managerial changes. Scientists expect this; after all this is how they work, so they would expect a research organisation to do the same and seek to acquire and base its decisions on evidence. For example, if the idea is to move everyone to open plan offices to encourage interaction, what is the evidence that this actually succeeds without reducing productivity?

Consider the value of any new administrative task you wish to impose

Recognise that if you want scientists to carry out new or more administrative tasks, there will be a cost in lost scientific outputs or grant proposals written. Consider whether the value of that new administrative task outweighs the loss in scientific performance?

For any new software package introduced for administration, such as in project management, the first question should be 'does it pass the keystroke test'? In other words, does it require fewer keystrokes to carry out the same task than were required by its predecessor software?

Language

Avoid business language. Scientists know that to make their work readily understood they must use clear and unambiguous language. In our country, phrases like 'going forward, cascading down, leveraging, holistic, roadmaps, platforms, benchmarks and decanting staff' are currently popular with managers. By using such phrases, instead of making their audience appreciate the professionalism and business awareness of their managers, the opposite effect is achieved. This is also true with the frequent use of acronyms. Language like 'following the advice of SISBI the RMT has spoken with the SMT to agree a new LTS-S' brings a barrier down between the high priests of management and the foot soldier scientists. Similarly, it may at first seem convenient to reduce the words in your PowerPoint talks to staff by peppering them with the acronyms that are the everyday currency of your management team. But then don't be surprised if you end up mystifying half or more of your audience! As they will be junior to you and won't ask for clarification, the presentation will have failed in its purpose, and so will the next one.

Communication and the role of email

It is now almost impossible to imagine the management of large scientific organisations and collaborative science projects without email. Many people work at a computer or have instant access to their email every hour of every day. For the busy manager juggling many tasks simultaneously, it is their most vital tool. However, emails can bring with them two extremely serious problems. The first is that their brevity and wording can cause misunderstanding, leading to upsets between people. The second is that they become a perpetual distraction. The manager can tend to rely on email rather than face to face communication with staff. The danger is one of increasing misunderstanding and mistrust on both sides. Very few people take the time to write perfectly worded emails where no misunderstanding or upset can arise. Even when they do, things can still go horribly wrong. Trying to work with e-mails constantly pinging their arrival is like trying to play chess with several wasps buzzing round your head. Given that people are human and are quick to see and take offence from emails, our advice would be:

- Make a serious effort to regularly talk to your colleagues face to face or by phone. It will save you time in the long run and clear up hundreds of misunderstandings. By talking to people, you can judge better how they are understanding and responding to your views and requests.

- Go the extra mile to make your email diplomatic, human and unambiguous. Read it once or twice before sending.

- When staff have carried out a task you requested by email, make sure you thank them.

- No matter how busy you are, do not ignore emails from members of your staff.

Encouraging a scientific culture

You want to educate, encourage and stimulate your scientists to develop and exercise their critical faculties. You want to show how you value good science and you want others to learn how to identify good science. This could take a number of forms, such as weekly seminars or journal clubs, or an annual showcase where every person/grouping or research project gets to give a talk. You want your scientists to be able to present their work (or in the case of a journal club, a paper they value) to friendly scrutiny. Ensure everyone chips in. But you must BE THERE YOURSELF!

Checklist for Chapter 16: Wider aspects of science management

1   Decide on your core principles of management and disseminate these. If you value scientific quality very highly, then tell everyone. Do not be sucked into micro-management.

2   Whether you know it or not, you set the atmosphere or culture for those who work for you. This will influence all parts of their outlook and behaviour at work.

3   Management starts with recruiting great people. Character is often more important than knowledge. You can teach knowledge but you cannot change someone's character.

4   Project management has three pillars for you to control; quality, time and budget. Quality is clearly important but all three pillars need your attention.

5   Don't get so mired in process that you lose sight of the purpose of the organisation!

6   Empower people by giving them your trust.

7   When the opportunity arises, don't hesitate to praise staff, but be genuine.

8   Be loyal to your staff in public.

9   As opportunities change, you may need to re-structure, but give your staff a good rationale. Review evidence from the past to find out what worked?

10   Communicate clearly; avoid, if not eliminate entirely, as much jargon and acronym use as possible.

11   Reduce your emails and put effort into seeing people face to face.

CHAPTER 17

Final thoughts

Life is uncertain and unpredictable. Nothing in this book will guarantee you a career that will reach the pinnacle of scientific achievement. But we can guarantee good science will be vital to humanity's future. So in some shape or form science will need you!

What good science is and doing good science is not obvious to most people. Indeed, most scientists starting out on their careers would struggle to find a training course on this most central of topics. Instead, most will hope to gain the principles of good science from their supervisors or science leaders. We are concerned that this reliance is not a secure route for the dissemination of good science, which was the major reason why we wrote this book. Some will be lucky and some unlucky. Hence, we offer here the distillation of 60 plus combined years of thinking about what makes good science, which we are attempting to transfer to you, the next generation of scientists who will be our replacements. As the reader will now know, there are no absolute truths in science; alternative views to ours will be available.

As you strive to become a better scientist you will be learning more than you think. Besides learning how to interpret and make the best use of your data, you will acquire a wide range of 'soft skills' that will be very useful to you in the future, whatever career path you choose. We have already discussed a few of these soft skills, such as operating with integrity, managing your time wisely, presenting information clearly, establishing working relationships and coping with setbacks. Working in research will inevitably also strengthen your ability to collaborate, to work in teams, to problem solve, to be resourceful and adaptable in the face of change. You should know that this collection of soft skills will make you attractive to many potential employers outside science.

Through thinking about the principles and practical advice we offer here, we are confident that you will be able to enjoy your scientific tasks and career more. Science will need you to challenge preconceived ideas so that falsehoods can be eliminated. You will now be part of the perpetual renewal process that is at the heart of good science. Setbacks and frustrations will be part and parcel of this journey, but they will be worthwhile as scientific knowledge grows for the betterment of our planet. Every piece of good science you do will make a difference. Taking artistic licence, we would describe good science as a thing of beauty. Now it is your turn: good luck!

Index

Note: Page numbers in **bold** indicate checklists.